T0271850

Symbolic Regression

Symbolic regression (SR) is one of the most powerful machine learning techniques that produces transparent models, searching the space of mathematical expressions for a model that represents the relationship between the predictors and the dependent variable without the need of taking assumptions about the model structure. Currently, the most prevalent learning algorithms for SR are based on genetic programming (GP), an evolutionary algorithm inspired from the well-known principles of natural selection. This book is an in-depth guide to GP for SR, discussing its advanced techniques, as well as examples of applications in science and engineering.

The basic idea of GP is to evolve a population of solution candidates in an iterative, generational manner, by repeated application of selection, crossover, mutation, and replacement, thus allowing the model structure, coefficients, and input variables to be searched simultaneously. Given that explainability and interpretability are key elements for integrating humans into the loop of learning in AI, increasing the capacity for data scientists to understand internal algorithmic processes and their resultant models has beneficial implications for the learning process as a whole.

This book represents a practical guide for industry professionals and students across a range of disciplines, particularly data science, engineering, and applied mathematics. Focused on state-of-the-art SR methods and providing ready-to-use recipes, this book is especially appealing to those working with empirical or semi-analytical models in science and engineering.

Gabriel Kronberger serves as professor at the University of Applied Sciences Upper Austria, Hagenberg. His research is focused on symbolic regression algorithms and their application in science and engineering. From 2018 until the end of 2022, he headed the Josef Ressel Center for Symbolic Regression where he developed symbolic regression methods for semi-analytical modeling to improve interpretability, trustworthiness, and extrapolation capabilities.

Bogdan Burlacu is Professor of Machine Learning and Data Science at the University of Applied Sciences Upper Austria, Hagenberg, and a member of the Heuristic and Evolutionary Algorithms Laboratory (HEAL). He has been an active researcher in the symbolic regression community for over a decade; his main area of expertise is the development of new symbolic regression algorithms and software.

Michael Kommenda is a senior researcher at the University of Applied Sciences Upper Austria, where he leads applied research projects with a focus on machine learning and data-based modeling. He authored several papers on symbolic regression and genetic programming and organized the workshop for symbolic regression at the Genetic and Evolutionary Computation Conference (GECCO).

Stephan Winkler serves as professor at University of Applied Sciences Upper Austria, Hagenberg; he is member of the Heuristic and Evolutionary Algorithms Laboratory (HEAL) and head of the Department of Medical and Bioinformatics as well as the Bioinformatics Research Group. For more than twenty years, he has been an active researcher in genetic programming and symbolic regression. Stephan Winkler has published numerous articles and books on data science and bioinformatics, and is member of the organization team of the Genetic Programming in Theory and Practice Workshop (GPTP).

Michael Affenzeller has published several papers, journal articles and books dealing with theoretical and practical aspects of evolutionary computation, genetic algorithms, and meta-heuristics in general. In 2001 he received his PhD in engineering sciences and in 2004 he received his habilitation in applied systems engineering, both from the Johannes Kepler University of Linz, Austria. Michael Affenzeller is professor for heuristic optimization and machine learning at the Upper Austria University of Applied Sciences, Campus Hagenberg, and head of the research group HEAL.

Symbolic Regression

Gabriel Kronberger, Bogdan Burlacu,
Michael Kommenda, Stephan M. Winkler, and
Michael Affenzeller

CRC Press
Taylor & Francis Group
Boca Raton London New York

CRC Press is an imprint of the
Taylor & Francis Group, an **informa** business

A CHAPMAN & HALL BOOK

First edition published 2025
by CRC Press
2385 NW Executive Center Drive, Suite 320, Boca Raton FL 33431

and by CRC Press
4 Park Square, Milton Park, Abingdon, Oxon, OX14 4RN

CRC Press is an imprint of Taylor & Francis Group, LLC

Library of Congress Cataloging-in-Publication Data
Names: Kronberger, Gabriel, author. | Burlacu, Bogdan, author. | Kommenda, Michael, author. | Winkler, Stephen M., author. | Affenzeller, Michael, author.
Title: Symbolic regression / Gabriel Kronberger, Bogdan Burlacu, Michael Kommenda, Stephen M. Winkler, Michael Affenzeller.
Description: First edition. | Boca Raton, FL : CRC Press, 2024. | Includes bibliographical references and index. | Summary: "Symbolic regression (SR) is one of the most powerful machine learning techniques that produces transparent models, searching the space of mathematical expressions for a model that represents the relationship between the predictors and the dependent variable without the need of taking assumptions about the model structure. Currently, the most prevalent learning algorithms for SR are based on genetic programming (GP), an evolutionary algorithm inspired from the well-known principles of natural selection."-- Provided by publisher.
Identifiers: LCCN 2024000819 (print) | LCCN 2024000820 (ebook) | ISBN 9781138054813 (hbk) | ISBN 9781032787053 (pbk) | ISBN 9781315166407 (ebk)
Subjects: LCSH: Genetic programming (Computer science) | Machine learning--Mathematics. | Mathematical models. | Logic, Symbolic and mathematical. | Regression analysis. | Engineering mathematics.
Classification: LCC QA76.623 .K76 2024 (print) | LCC QA76.623 (ebook) | DDC 006.3/823--dc23/eng/20240118
LC record available at https://lccn.loc.gov/2024000819
LC ebook record available at https://lccn.loc.gov/2024000820

ISBN: 978-1-138-05481-3 (hbk)
ISBN: 978-1-032-78705-3 (pbk)
ISBN: 978-1-315-16640-7 (ebk)

DOI: 10.1201/9781315166407

Typeset in Nimbus Roman
by KnowledgeWorks Global Ltd.

Publisher's note: This book has been prepared from camera-ready copy provided by the authors.

Contents

Preface

More than three decades ago, John Koza coined the term "Symbolic Regression" in his book series with which he popularized genetic programming, a nature-inspired method to evolve programs to solve algorithmic tasks. Since then symbolic regression has become almost the flagship-application for genetic programming. This goes so far that some researchers even started to use genetic programming (GP) and symbolic regression (SR) interchangeably. However, we think SR – which describes the idea of using a symbolic representation for nonlinear regression in combination with symbolic operations for fitting the model – is an important concept that is worthwhile to be treated independently, not least because of its capability to produce interpretable models. In our point of view, GP is one of many possible approaches to solve SR tasks, albeit the most popular. Consequently, we also mainly treat GP in this book. Since the introduction of GP and SR in the early 1990s, many alternative methods for SR have been developed. Especially in the last ten years SR has become more popular outside of the evolutionary computation community in different areas of science and engineering. The increasing interest in SR has been driven mainly by its potential to identify interpretable models – an important aspect in the context of explainable artificial intelligence and scientific machine learning.

We feel there is a need for a new book dedicated exclusively to SR. Koza's books on GP have many SR examples, but they do not cover more recent developments. Several books – including our earlier book with CRC Press *Genetic Algorithms and Genetic Programming* – contain chapters or a few sections dedicated to SR, but do not give a comprehensive treatise of the topic. These facts, and the increased popularity and usefulness of SR have motivated us to prepare this book.

SR has been one of our major research interests for more than twenty years. In the research group *Heuristic and Evolutionary Algorithms Laboratory* (HEAL) of the University of Applied Sciences Upper Austria, Hagenberg Campus we have accumulated a lot of experience on using GP and SR for data-based modelling in various projects, often together with R&D-heavy companies in the region of Upper Austria. Michael Affenzeller, the head of the research group, and Stephan Winkler, at that time PhD student of Michael Affenzeller, first experimented with GP and SR in 2004 and implemented tree-based GP for the first version of HeuristicLab, an open-source software framework developed and used within HEAL. Stephan Winkler's dissertation on *Evolutionary System Identification: Modern Concepts and Practical Applications* was mainly on GP and SR, and since that time he has continued to do research in this area. Since

2020 he has been acting as one of the organizers of the yearly workshop on *Genetic Programming Theory and Practice* held at the University of Michigan or Michigan State University.

The work by Michael and Stephan sparked several follow-up projects and growth of the team. In 2005, Gabriel Kronberger joined HEAL and started to work on GP and SR together with Michael and Stephan. Michael Kommenda, who joined the group in 2008, and Gabriel created a new implementation of tree-based GP for the third major release of HeuristicLab. Several algorithmic improvements and functionality to make model analysis and validation more convenient were developed by Gabriel and Michael while working on their dissertations. Bogdan Burlacu joined the team in 2011 and worked on methods for analysis of diversity, schemas, and heredity to improve the understanding of GP. He also developed several extensions to the GP code in HeuristicLab and later implemented and published Operon, a high-performance implementation of tree-based GP for SR in C++. Operon is a reimplementation of the most relevant parts of the SR implementation from HeuristicLab with a focus on performance. The three dissertations *Symbolic Regression for Knowledge Discovery: Bloat, Overfitting, and Variable Interaction Networks* by Gabriel, *Tracing of Evolutionary Search Trajectories in Complex Hypothesis Spaces* by Bogdan, and *Local Optimization and Complexity Control for Symbolic Regression* by Michael describe several of the SR developments and applications that we worked on in the time frame from 2008 – 2018. Michael Affenzeller was the first supervisor for all of these doctoral theses.

We had four goals for the book. First, it should be accessible to researchers and practitioners in science and engineering without a background in algorithms and evolutionary computation. There are plenty of publications mainly aimed at researchers with a background in machine learning. We instead decided to use a hands-on approach targeted towards an applications-oriented audience, because we believe that SR is a hidden champion that can benefit practitioners in many domains.

Our second goal was to provide many real-world examples for SR to demonstrate that SR is not just an academic idea. There are so many interesting applications in various areas of science and engineering where SR can be an ideal modelling approach and we wanted to provide a demonstration for this wide applicability. A longer book is the perfect opportunity to discuss different applications in detail and to give step-by-step instructions for real-world use cases, because we do not have to focus on computer experiments and comparative analysis which would be required for a journal or conference article. We decided to skip such comparative analysis completely in this book, because our goal is not to show that a certain SR algorithm produces better results than method A or method B. Instead, our goal is to demonstrate the capabilities of SR while also discussing potential issues that might occur. A detailed discussion of many different examples may initiate new ideas how a similar approach could be used in another context and might lead some of our readers to try SR in related scenarios.

Our third goal was to describe in detail how to solve actual problems with SR instead of describing in detail how SR algorithms work. This goal is directly related to our decision to write this book for practitioners. We wanted to provide clear, almost tutorial-like descriptions for successful applications of SR. For example, a common difficulty for practitioners is how to choose algorithm parameters. Most GP implementations for SR have many parameters that can only be adjusted appropriately with a fundamental understanding of GP. This can be daunting for engineers that are used to other regression methods where only a few parameters are relevant. To help practitioners, we could have written detailed tutorials specifically for a single SR software tool, for instance using our own software HeuristicLab or Operon. However, software systems change quickly which means that a book about a certain software system would be outdated soon.

This led to our fourth goal which is that the book should discuss SR generally without a focus on a specific software tool. Instead we describe the general principles and give examples of applications to guide our readers in their own efforts. We include a short section where we list software systems and online resources for SR.

The book consists of two larger parts; the first part is focused on SR methods and the second part on applications. In the first part we start with a brief introduction of machine learning for data-based modelling, and then give a detailed description of the mechanics of GP. Practitioners should certainly read the first chapter but may want to skip the chapter on evolutionary computation on the first reading. Several sections in that chaper discuss concepts in depth and can be interesting for readers that want to gain a deeper understanding of evolutionary computation methods.

We dedicate a large chapter to GP and the fundamentals of evolutionary computation because GP has the longest history and is so far the best tested and studied method for SR. Even though different non-evolutionary solution methods have been described more recently, GP still works very well for many datasets. Evolutionary methods are inherently parallel and can therefore be easily parallelized to multiple cores or distributed to multiple nodes on a larger cluster. However, for many SR tasks that are practically relevant, a common office computer is sufficient to run GP.

A larger chapter that is relevant for all practitioners is devoted to SR model analysis, validation, and selection. This chapter contains specific advice for SR models and discusses several techniques to aid model selection and interpretation.

The second part of the book consists of a large chapter that is devoted to different application examples. Each section in this chapter describes a different application and can be read independently. We invite our readers to read the sections in any order and skip those examples that are less relevant to their own work.

We would not have been able to prepare this book without the support of many of our friends and colleagues, who came up with ideas that found their way into the book, provided success stories about innovative applications of SR, implemented software, proofread parts of this book, or supported us in other ways directly or indirectly. We thank our long-term collaborators working at the companies that have supported our research financially for many years. In alphabetical order we thank Deaglan J. Barlett, Andreas Beham, Bruno Buchberger, Luigi Del Re, Harry Desmond, Christoph Feilmayr, Pedro Ferreira, Fabricio Olivetti de Franca, Jose Ignacio Hidalgo, Witold Jacak, Evgeniya Kabliman, Lukas Kammerer, Johannes Karder, Berthold Kerschbaumer, E. Peter Klement, Michael Kordon, Johannes Kronsteiner, Bernhard Löw-Baselli, Falk Nickel, Sophie Pachner, Andreas Promberger, Stefan Scheidel, Stefan Wagner, and all those who we forgot to include in this list.

Gabriel Kronberger, Bogdan Burlacu, Michael Kommenda, Stephan M. Winker, Michael Affenzeller, Hagenberg, 2023

Symbols and Notation

$\boldsymbol{X}$	Matrix of input values with n rows of d real values. $= (x_{i,j})_{i=1..n, j=1..d}$
$\boldsymbol{x}$	Vector of input variable values.
$\boldsymbol{y}$	Vector of target variable values.
ε	Residual vector.
$|x|$	Absolute value of scalar value x.
$\|\cdot\|_p$	p-norm of vector or function.
$\boldsymbol{X}^\top$	Transpose of matrix $\boldsymbol{X}$.
$\boldsymbol{Xy}$	Matrix vector product of $\boldsymbol{X}$ and $\boldsymbol{y}$.
$f(\boldsymbol{x}, \boldsymbol{\theta})$	Function of variables $\boldsymbol{x}$ and parameter vector $\boldsymbol{\theta}$.
$\boldsymbol{X}^{-1}$	Inverse of matrix $\boldsymbol{X}$.
$\mathrm{diag}(\boldsymbol{x})$	Diagonal matrix with elements as given in vector $\boldsymbol{x}$.
$\bar{x}$	Arithmetic mean of the vector $\boldsymbol{x}$.
$\sigma(\boldsymbol{x})$	Standard deviation of vector $\boldsymbol{x}$.
$\mathrm{var}(\boldsymbol{x})$	Variance of vector $\boldsymbol{x}$ $(\mathrm{var}(\boldsymbol{x}) = \sigma(\boldsymbol{x})^2)$.
$\mathrm{cov}(\boldsymbol{x}, \boldsymbol{y})$	Covariance between vectors $\boldsymbol{x}$ and $\boldsymbol{y}$.
$\rho(\boldsymbol{x}, \boldsymbol{y})$	Pearson's correlation between vectors $\boldsymbol{x}$ and $\boldsymbol{y}$.
R^2	Coefficient of determination $R^2 = 1 - \frac{1}{\mathrm{var}(\boldsymbol{y})} \frac{1}{n} \|\boldsymbol{y} - f(\boldsymbol{X})\|_2^2$.
$\mathscr{L}(\boldsymbol{\theta})$	Likelihood function for given observations and a parametric model $f(\boldsymbol{\theta})$.
$\mathscr{L}(\boldsymbol{\theta}; D)$	Likelihood function, where the dependency on a dataset D is explicitly noted.
$\dot{y} = \frac{\mathrm{d}y}{\mathrm{d}t}$	First derivative of $y(t)$ in an ordinary differential equation. For brevity we do not explicitly note the dependency $y(t)$.
$\ddot{y} = \frac{\mathrm{d}^2 y}{\mathrm{d}t^2}$	Second derivative of $y(t)$.
$\frac{\partial y(\boldsymbol{x})}{\partial x_i}$	Partial derivative of function $y(\boldsymbol{x})$ over i-th variable x_i.
$\boldsymbol{\eta}$	Vector of hyperparameters, usually algorithm parameters.
$\mathscr{F}$	Function set.
$\mathscr{T}$	Terminal set.
$\mathscr{P}$	Primitive set $(= \mathscr{F} \cup \mathscr{T})$.
$\Re(z)$	Real part of complex value z.
$x \sim N(\mu, \sigma)$	x is sampled from a normal distribution with mean μ and standard deviation σ.
$x \sim U(l, h)$	x is sampled from a uniform distribution and $\in [l, h]$.
Ω	Limited d-dimensional input space $\subset \mathbb{R}^d$.

1

Introduction

Symbolic regression (SR) searches the space of mathematical expressions for a model that represents the relationship between the predictors and the dependent variable. The currently most common and also most successful learning algorithms for SR are based on genetic programming (GP), an evolutionary algorithm inspired from the well-known principles of natural selection. The basic idea of GP is to evolve a population of solution candidates in an iterative, generational manner, by repeated application of selection, crossover, mutation and replacement. Thanks to this approach, the model structure, coefficients, and input variables are searched simultaneously without the need to make assumptions. The initialization of the first generation is usually done randomly so as to generate advantageous initial diversity. The user can influence the characteristics of the search space by specifying the allowed mathematical operations or the allowed size of the models. From one generation of solution candidates new candidates emerge by crossing pairs of above-average individuals, thus ensuring that each new candidate inherits a mixture of properties of its parents. To introduce new genetic diversity, the child individual created via crossover is mutated with a certain probability. The process of replacement is responsible for determining which children will become parents for the next generation. This iterative process is repeated over many generations until a termination criterion is met and usually the best solution candidate from the last generation is returned.

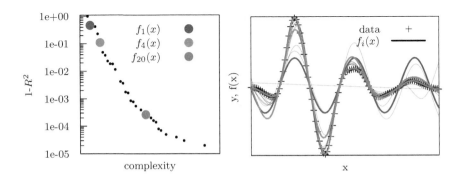

Figure 1.1: Data and predictions for three solution candidates (Equations (1.1) to (1.3)) with different tradeoff between goodness-of-fit and size.

$$f_1(x) = \sin(1.992\,x) \tag{1.1}$$

$$f_4(x) = \exp(\sin(0.526\,x))\sin(1.995\,x) \tag{1.2}$$

$$f_{20}(x) = \sin(0.307\,x)\sin(2\,x)$$

$$\exp\Big(\sin(0.637\,x) - \cos(\sin(\sin(0.485\,x)) - 1.341) \tag{1.3}$$

$$\cos(0.461\,x)(\sin(2.082\,x) + 0.123\,x)\Big)$$

Figure 1.1 shows an exemplary result of a run of genetic programming for SR for a highly nonlinear dataset. Many GP systems produce a set of solutions from which the user can select one or multiple alternatives based on his or her preference for goodness-of-fit and model size. For this dataset it is helpful to include trigonometric functions (*sin, cos*) to capture the periodic behaviour. Genetic programming is able to assemble the elements from the function set to form short expressions that fit the data. Since the dataset in this example is small, it should only take a few seconds to find acceptable solutions. The left plot shows the predictions of multiple models with the best accuracy and length (Pareto front). Except for the extremely short models, most of the models fit the data well. The predictions of three selected models with different length/accuracy tradeoffs are highlighted.

The aspects of interpretability and explainability are increasingly gaining in importance and perception in artificial intelligence (AI) and machine learning (ML). A common criticism of ML methods is that these approaches can only identify correlations in the data, but not causality. If, however, it is possible to provide models that can be explained and interpreted, this opens up the possibility of specifically questioning the causality of models learned from data with the aid of domain expertise. Nowadays, we observe various efforts to make black box methods such as deep neural networks more interpretable, which seems inherently difficult and only possible to some extent (Rudin, 2019). Alternatively, SR makes it possible to learn complex nonlinear system behaviour from data in such a way that the produced models, by virtue of their transparent structure, can be explained and interpreted by experts.

Explainability and interpretability are key elements for integrating humans into the loop of learning in AI. Nowadays, learning in AI is mainly associated with ML from data. If it becomes increasingly possible for the data science expert and domain expert to better understand the internal processes of the algorithms and to better interpret the resulting models, they can participate in the process of learning more holistically. With the help of better explainable and interpretable algorithms and models, learning can become increasingly possible in all directions and strengthens the interconnection between data, humans and machines. Deductive and inductive approaches should no longer be an either/or decision, but should complement each other synergistically. Scientific and technical knowledge from the domain will thus be able to flow into data-based models just as much as human experience or even intuition.

This applies not only to the interpretation and selection of the final results, but already during the search, where humans will be able to influence which solution properties they prefer. In a next step, the machine can systematically learn via interacting with humans and systematically make the knowledge gained available. SR, with its unique ability to enable accurate modelling of complex nonlinear relationships in interpretable and explainable terms, is thus, in our view, a key technology toward human-centered AI. SR occupies a special place among ML methods due to its ability to identify complex nonlinear relationships without making prior assumptions neither about the structure of the model nor about its coefficients. Another unique feature of SR is that, when applied to appropriate problems, it does not require a tradeoff between interpretability and model accuracy. A preconception often heard nowadays is that one has the choice either to generate accurate models from a lot of data using techniques like deep neural networks or to be satisfied with simple models like linear regression models. We have made the observation in many primarily industrial research collaborations that models created with SR are not only interpretable but can also produce very accurate predictions for highly nonlinear systems (La Cava et al., 2021). This is especially successful if the background of the systems to be modeled is technical or mechatronic, or more general, in which the description language of mathematics is the natural one.

Data from many natural and technical fields, abundantly available today in numerical form from simulations and sensor measurements, is ideally suited for SR. If it is possible to capture the system characteristics by a mathematical model, not only very good results regarding interpolation are possible, but also regarding extrapolation. The possibility of specifying the permitted mathematical functions and linking rules allows a definition of the search/hypothesis space tailored to the concrete task.

The complexity and nonlinearity of the evolved mathematical models can be tailored to the task by the user, for example through the definition of the set of allowed functions and operators. With background knowledge from the domain or from previous modelling attempts, the user can determine whether a rather small set consisting only of addition and multiplication is sufficient, or whether it is perhaps better to also include logarithms, exponential functions, and/or trigonometric functions. Allowing conditional functions and Boolean functions often makes sense for classification problems where models in the form of decision trees are often useful. Especially the representation of a hypothesis as a mathematical model opens comparatively easy hybridization possibilities of knowledge learned from data with prior knowledge from the respective domain. Such fine control of the complexity of the function and the hypothesis space is unique among ML methods.

This book represents a practical guide for industry professionals and students in the fields of data science, engineering, applied mathematics, and several others. The intended audience are people working with empirical or semi-physical data-based models for process control and optimization. This book is focused on applications of modern state-of-the-art methods for SR and provides a variety of examples and ready-to-use recipes for obtaining, selecting and analyzing data models in different application scenarios. The material is accessible for readers with an undergraduate level of mathematics. Readers of the book should ideally have background knowledge in the area of empirical data-based modelling or statistical learning, but this is not a prerequisite. Throughout the book we often mention related work only briefly so that we do not interrupt readers with side information which might be relevant to only a few of our readers. In such cases we give references to relevant publications with more detailed information at the end of each chapter.

In a few places throughout the book we link directly to online resources but we use such links sparingly. Additional material including datasets and example scripts as well as up-to-date links to SR software tools and other software resources are available on the book website `https://symreg.at/book`.

We start with a brief overview of the basic knowledge required in the field of ML in Chapter 2. Chapter 3 already focuses on SR; evolutionary computation and especially genetic programming as a hypothesis space search method for SR are described in Chapter 4. Chapter 5 deals with several refinements of the basic algorithm dedicated to the validation, inspection, simplification, and selection of models, followed by Chapter 6 which summarizes advanced techniques such as the integration of knowledge, optimization of coefficients, and other extensions of the standard workflow. Chapter 7 looks at representative examples and applications of SR in different areas of regression, classification, and time series analysis; both well-known benchmark datasets from the scientific literature and real-world datasets are used here. Chapter 8, finally, concludes this book. Additionally, software packages implementing SR and relevant benchmarks are summarized in the appendix.

2

Basics of Supervised Learning

2.1 Introduction

Machine learning (ML) is the branch of computer science that studies the development of methods and algorithms that can learn how to describe a system or how to perform a particular task from data. Depending on how the data is organized and structured, ML can be divided in two main subfields, called supervised and unsupervised learning.

If the sample data is labeled, meaning that each data point is already mapped to a corresponding measurement known to be correct, then we are dealing with a supervised learning problem, where the goal is to develop a predictive model that can predict a system's future behaviour from past measurements or observations. This learning process is also called fitting or training which refers to adapting model parameters so that the model matches observations. Classification, regression, and time series modelling are the principal examples of supervised learning tasks (Figure 2.1).

If the data is unlabeled then we are dealing with an *unsupervised* learning problem, where the goal is to discover hidden patterns in data. Clustering, dimensionality reduction, or association rule mining are main examples of unsupervised learning tasks.

ML is often used in the same context as data mining (Han et al., 2011), which has a good reason in that one cannot do without the other: while ML techniques are used to learn models from data, data mining is rather concerned with finding patterns and correlations in data. Data science (Schutt and O'Neil, 2013) is even more general, it is an interdisciplinary field of computer science that focuses on methods and algorithms for extracting knowledge from data, both structured as well as unstructured.

Condensed to one word, we can say that ML is the "how" while data mining is the "what". Through extended interpretability and explainability as supported with symbolic regression, the focus on the "why" can and should be further intensified.

Symbolic regression (SR) belongs to the family of supervised learning algorithms for regression problems. The main appeal of SR is that it can assemble a predictive model from scratch, starting only from a given set of operators and operands and a basic set of rules for combining them. This means that no prior knowledge is required to identify interpretable models.

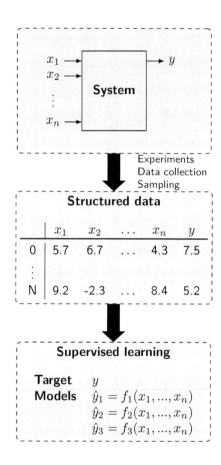

Figure 2.1: Data-based system identification relies on data that are collected via experiments. These data are then the basis for ML that identifies models and formulates relationships.

This concept may also be used for classification to learn a discriminant function which is used together with threshold values to produce discrete classes.

We give an overview of the main approaches for regression analysis, in preparation for the next chapter where we discuss specific topics pertaining to SR.

2.2 Regression

With a model-based approach a regression problem is defined as the task of finding a model for a dependent variable y as a function of a set of independent variables x_i and their associated parameters. In this context y is also called the *target* or *dependent* variable while x_i are called *predictors* or *independent* variables.

The problem can be expressed as

$$y = f(X, \theta) + \varepsilon$$

where $X = [x_1, ..., x_n]$ is the matrix of input vectors, θ is a parameter vector associated with the model, and ε is an additive error term that subsumes random statistical noise or other irreducible error sources.

The goal of regression analysis is to determine the function $f(x, \theta)$ that accurately explains the data. Depending on whether f is linear or not in the parameters θ_i, we distinguish between *linear* and *nonlinear* regression, with the former being simpler than the latter and often used as an accuracy baseline.

2.2.1 Linear Models

We assume a dataset consisting of n observations containing p input variables. The linear relationship between variables (also called *predictors*) and the dependent variable can then be expressed in matrix form:

$$y = X\beta + \varepsilon$$

where

$$y = \begin{pmatrix} y_1 & y_2 & \cdots & y_n \end{pmatrix}^\top$$

$$X = \begin{pmatrix} \mathbf{1}^\top & \mathbf{x}_1^\top & \mathbf{x}_2^\top & \cdots & \mathbf{x}_n^\top \end{pmatrix} = \begin{pmatrix} 1 & x_{11} & \cdots & x_{1p} \\ 1 & x_{21} & \cdots & x_{2p} \\ \vdots & \vdots & \ddots & \vdots \\ 1 & x_{n1} & \cdots & x_{np} \end{pmatrix}$$

$$\beta = \begin{pmatrix} \beta_0 & \beta_1 & \cdots & \beta_p \end{pmatrix}^\top$$

$$\varepsilon = \begin{pmatrix} \varepsilon_1 & \varepsilon_2 & \cdots & \varepsilon_n \end{pmatrix}^\top$$

The parameter vector $\boldsymbol{\beta}$ includes an additional intercept term β_0 corresponding to the constant column $x_{i0}, i = 1, \ldots, n$ in $\boldsymbol{X}$. It is important to observe here that the linearity assumption refers to the regression coefficients β_i and not to the predictor variables x_i which can occur multiple times in the expression, with different transformations (e.g., x_1, x_1^2, and so on). For example, polynomial regression generates a model of the target nonlinear in the predictors x_i but linear in the parameters β_i. The vector $\boldsymbol{\varepsilon}$ captures disturbances or noise that have an impact on the target y.

Due to its simplicity, linear regression is often used as an initial exploratory step to check if the studied process or system can be explained by a linear model or to establish an accuracy baseline.

2.2.2   Nonlinear Models

In nonlinear regression we assume a model function nonlinear in the components of the vector $\boldsymbol{\theta}$, of the form:

$$\boldsymbol{y} = f(\boldsymbol{X}, \boldsymbol{\theta}) + \boldsymbol{\varepsilon}$$

Depending on initial assumptions about the model, we distinguish between two main approaches for solving nonlinear regression:

Deductive or *top-down* modelling approaches assume some kind of relationship between the target and input variables, for which the structure is known but the parameters are unknown. The identification of parameter values to minimize $\boldsymbol{\varepsilon}$ is then an optimization problem that can be solved – depending on the structure of the model and the error measure to be minimized – with iterative methods such as different forms of gradient-based methods (Bubeck, 2015; Nesterov, 2003), derivative free methods (e.g., simulated annealing (Pincus, 1970), or evolution strategies (Rechenberg, 1973)).

Inductive or *bottom-up* modelling does not assume any relationship between the input variables and the target, and attempts to identify the function f by building a model that minimizes the error measure. Apart from SR, which is discussed in further detail in the next chapters, some established inductive learning methods for solving nonlinear regression are:

• Random Forests (RF) (Breiman, 2001) exploit the idea of bagging multiple models to increase the quality of the prediction. The base learner in a RF is a binary decision tree. RF generate predictions from ensembles of decision trees, where the predicted value is given by majority vote (classification) or mean response (regression).

• Gradient Boosted Trees (GBT) (Friedman, 1991; Chen and Guestrin, 2016) are similar to RF. However, the set of regression trees is learned iteratively, whereby the residuals left from earlier iterations are used as the target for the next iteration. In contrast to RF, the trees built with gradient boosting are small. The final model typically consists of thousands of trees.

- Support vector machines (SVM) (Vapnik, 1998) use the kernel-trick by projecting the input to a high-dimensional space where the samples are linearly separable.

- Gaussian process regression (GPR) (Rasmussen and Williams, 2006) like SVM uses the kernel trick to learn a linear model in the projected high-dimensional space. It is a Bayesian model and learns prediction intervals for the identified nonlinear model.

- Artificial neural networks (ANN) (Bishop, 1995) and deep neural networks (Bengio and LeCun, 2007), (Goodfellow et al., 2016). ANN consist of neurons organized in a hierarchical collection of layers such that activation paths are learned by the network through the tuning of numerical weights that control the strength of each neuron's signal.

ANNs, SVMs, and GPR models assume an initial structure (i.e., the topology of the layers for ANN or the type of kernel and its hyper-parameters for SVM and GPR) which strongly influences the characteristics of functions that can be expressed with the models. Their success therefore depends strongly on choosing the correct initial structure. However, the function identified through fitting the model is mainly determined by the parameters of the model and only indirectly by the structure or kernel. As it is difficult to understand the effects of the parameter values on outputs of these models, each of these models is used as a *black box*.

RF and GBT are a little different, in that they do return an inspectable model structure, as a collection of decision trees in which every tree node is split into *True* and *False* branches according to a question asked about the value of a feature. However, the final model is made up of hundreds or thousands of decision trees which can be very deep in the case of RF, therefore the internals are not easily understandable and just like above, the model is considered a black box.

In both cases, it can be interesting to look inside the black box and calculate statistics over the parameters or trees in the model. For instance the variables used more often in the first decision nodes in RF and GBT models are more relevant to predict the target variable than variables that are seldomly used.

2.2.3 Error Measures

Error measures are used in supervised learning to assess the quality of the predictive model. In the context of mathematical optimization which is at the core of supervised learning, we use the error measure as the *objective function* or *loss function* to be minimized. The used error measure influences the modelling results, as different error measures induce different optimal models. We choose the error measure based on the requirements of the application. Whereby, we concern ourselves mainly with following aspects:

- Is the measure absolute or relative?

- Does it measure error or correlation?

- Is it sensitive to outliers?

- Is it informative for small as well as large data ranges?

Many common error measures are functions of the *error* or *residual vector* $\varepsilon = f(\boldsymbol{X}, \boldsymbol{\beta}) - \boldsymbol{y}$. The most commonly used error measure in SR literature is the *residual sum of squares* (RSS), also known as the *sum of squared residuals* (SSR) or the *sum of squared errors* (SSE):

$$\text{SSE} = \sum_{i=1}^{n} \varepsilon_i^2 = \|\varepsilon\|_2^2 \tag{2.1}$$

The SSE depends on the number of observations n so we cannot compare SSE values directly for datasets with different sizes. The *mean of squared errors* (MSE) over the observations is defined as

$$\text{MSE} = \frac{\text{SSE}}{n}$$

and allows such comparisons.

The SSE and the MSE are absolute measures and sensitive to outliers. An absolute error measure that is less sensitive to outliers than the MSE is the *mean of absolute errors* (MAE).

$$\text{MAE} = \frac{1}{n} \sum_{i=1}^{n} |\varepsilon_i| = \frac{\|\varepsilon\|_1}{n}$$

This error measure can be used for *robust regression* to limit the effect of outliers caused e.g., by incorrect measurements or data entry.

Relative measures capture the contribution of errors relative to the target variable values. We frequently use the normalized MSE (NMSE) as an error measure in the later chapters which is related to the $R^2 = 1 - \text{NMSE}$. It puts the MSE in relation to the variance of the target variable and enables comparisons over different datasets with target variables with different variances. The NMSE is in the range $[0\% \ldots 100\%]$ for correctly scaled models. It is easy to show that a model with an NMSE larger than one has a higher MSE than the constant mean model which outputs the arithmetic average of $\boldsymbol{y}$ for all inputs.

$$\text{NMSE} = \text{var}(\boldsymbol{y})^{-1} \text{MSE}$$

An alternative to the NMSE is the *coefficient of determination* R^2 that captures the proportion of the variation in the target variable which is explained by the predictive model. The R^2 and the NMSE are directly related and can be inferred from each other.

$$R^2 = 1 - \text{SSE} \left(\sum_{i=1}^{n} (y_i - \overline{y})^2 \right)^{-1} = 1 - \frac{\text{MSE}}{\text{var}(y)} = 1 - \text{NMSE}$$

The mean absolute relative error (MARE)

$$\text{MARE} = \frac{1}{n} \sum_{i=1}^{n} |\varepsilon_i y_i^{-1}|$$

weights residuals by the corresponding target value and therefore allows larger absolute errors for larger absolute target values while target values close to zero must be predicted more accurately. The MARE is only defined for target vectors without zeros and is a good choice when the error of the measurement correlates with the measured value meaning the measurement is less accurate for larger absolute values. Other well known loss functions for regression include the epsilon-insensitive loss function (Dekel et al., 2005) from support vector regression or the Huber loss (Huber, 1964).

The concept of loss functions or objective functions is generic and used for all types of different mathematical optimization problems of which supervised learning tasks including regression and classification are special cases. Regression and classification tasks are typically approached with statistical methods. In this context, the likelihood function can be used to calculate the likelihood of the data given the model and we try to find the maximum likelihood model. The likelihood function is a direct result of model assumptions, most importantly the distribution of errors. An important example is the likelihood function for independent and identically distributed Gaussian errors ($y_i = f(x_i) + \varepsilon_i, \varepsilon_i \sim_{\text{i.i.d}} N(0, \sigma_{\text{err}})$),

$$\mathscr{L}(\theta, \sigma_{\text{err}}) = \prod_{i=1}^{n} \frac{1}{\sqrt{2\pi\sigma_{\text{err}}^2}} \exp\left(-\frac{(y_i - f(x_i, \theta))^2}{2\,\sigma_{\text{err}}^2} \right)$$

It is easy to show that minimization of SSE is equivalent to maximization of this likelihood and therefore implicitly assumes identically distributed errors. Standard errors for each observation can be easily incorporated into the likelihood function or by using weighed squared errors. Similarly, minimization of MAE is equivalent to maximization of the likelihood function for Laplace distributed errors.

2.3 Classification

In contrast to regression where the target variable has values on a continuous scale, in classification the target variable is discrete, whereby each observation is assigned a class label. The task of classification can also be accomplished using regression methods, by converting the continuous numerical output of a regression model into a choice. For instance, logistic regression solves classification tasks by learning a model that produces values in the unit interval [0..1] using the nonlinear logistic function on top of a linear model. The output of the logistic regression model can be interpreted as a probability for the positive class.

If the class labels for the dependent variable are binary (such as positive/negative, true/false, yes/no, 0/1) then we are dealing with a *binary* classification problem. If we have $k \geq 3$ class labels we are dealing with a *multiclass* (or *multinomial*) classification problem.

Error measures for classification models are typically defined in terms of the ratio of correctly classified samples. The standard nomenclature as shown in Figure 2.2 is commonly used (for conciseness, a binary classification problem is considered). *Positives* (P) is the number of positive samples. *Negatives* (N) is the number of negative samples. *True positives* (TP) is the number of positive samples (P) also predicted to be positive, while *true negatives* (TN) is the number of negative samples (N) also predicted to be negative. In contrast, *false positives* (FP) is the number of negative samples (N) wrongly predicted to be positive; these samples can be interpreted as false hits and are called *type I errors*. *False negatives* (FN) is the number of positive samples wrongly predicted to be negative; these samples can be interpreted as missed hits and are called *type II errors*.

On the basis of these quantities several measures are defined. *Accuracy (ACC)* represents the ratio of correctly classified samples

$$\text{ACC} = \frac{\text{TP} + \text{TN}}{\text{P} + \text{N}} \tag{2.2}$$

Sensitivity is the ratio of positive samples that are correctly classified, also called *true positive rate (TPR)* or *recall*

$$\text{TPR} = \frac{\text{TP}}{\text{P}} \tag{2.3}$$

Specificity represents the ratio of negative samples that are correctly classified, also called *true negative rate (TNR)*

$$\text{TNR} = \frac{\text{TN}}{\text{N}} \tag{2.4}$$

Precision is the ratio of positive predictions that are correct, also called *positive predictive value (PPV)*

$$\text{PPV} = \frac{\text{TP}}{\text{TP} + \text{FP}} \tag{2.5}$$

False discovery rate (FDR): the ratio of positive predictions that are incorrect

$$\text{FDR} = \frac{\text{FP}}{\text{FP} + \text{TP}} = 1 - \text{PPV} \tag{2.6}$$

The F_1 *score* (also called F-score or F-measure) considers both the precision and the recall of a classifier, and is calculated as the harmonic mean of precision and recall:

$$F_1 = 2\frac{PPV \cdot TPR}{PPV + TPR} \qquad (2.7)$$

These measures are particularly useful when arranged in a confusion matrix as shown in Figure 2.2 and can offer an intuitive picture of classifier performance together with the Receiver-Operating-Characteristic (Fawcett, 2006).

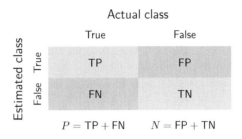

Figure 2.2: Schema of a binary confusion matrix showing P, N, TP, TN, FN, and FP.

2.4 Time Series Prediction

If the analyzed system's behaviour is to be modeled not only considering current data but also considering how the system changes over time (dynamics), then our goal is to find a time series model. There are many different variants of time series models (see for example Montgomery et al. (2007)). Based on what we have already discussed above, a straightforward extension is to include lagged variables. Here a lagged variable means that we use values that a variable had at an earlier point in time. Such a model could have the form

$$
\begin{aligned}
y(t) = f\Big(&\boldsymbol{x}(t), \boldsymbol{x}(t-1), \boldsymbol{x}(t-2), \dots, \boldsymbol{x}(t-o_{max}), \\
&y(t-1), y(t-2), \dots, y(t-o_{max}), \boldsymbol{\theta} \Big) + \varepsilon(t)
\end{aligned}
\qquad (2.8)
$$

where o_{max} defines the maximum offset which may be selected based on prior knowledge about the system.

This is an example of an *autoregressive* model which means that lagged values of the target variable are used as input variables. Predictions over multiple time steps with such models require feeding predicted values back into the model as inputs. Here we have to be aware that future values of the inputs $x(t + h)$ must be known up to the prediction horizon $t + h$. Additionally, prediction errors of the model may be amplified when feeding back the predicted values for y into autoregressive models. This can be problematic for larger prediction horizons, as future values of inputs might be known, but future values of the target variables can only be estimated.

For time series modelling we need a database storing measurements for the system that were recorded over time; these data should ideally be available at equidistant time stamps. Then we can fit models for time series using any of the previously mentioned regression methods with lagged variables. When the measurements are not equidistant we may interpolate values or alternatively use a differential equation to represent the model, instead of the simple autoregressive form above. This is described in more detail in Chapter 6.

A simple way to learn regression or classification models for time series that works with all software systems is to add lagged copies of the input variables to the database. For example, if the variable x_2 is only allowed as input with a time offset, then we remove $x_2(t)$ from the database and add its lagged copies $x_2(t-1), x_2(t-2), \ldots, x_2(t - o_{max})$. This approach is on the one hand easy to implement, but on the other hand it also increases the size of the database and – which is even worse – also increases the search space and increases the danger of overfitting. Models trained with this method ignore that $x_2(t-1)$ and $x_2(t-2)$ are often highly correlated features.

The error measures used for time series are mainly the same as for static regression and classification tasks. For time series with irregularly spaced measurements, the duration between measurements should be used to weight errors, to take into account that any prediction error accumulates up to the next available measurement.

Once again, SR offers alternative ways for time series modelling. In the next sections and chapters we will discuss methods that are able to learn the optimal time offset for each variable and search for models with arbitrary forms and time offsets, e.g., $y(t) = \log_2(x_5(t-2)) + 0.05\,x_4(t-1)$.

2.5   Model Selection

The main goal of ML is to produce models with the best performance on yet unobserved data collected after the training phase. Such models are said to generalize well. In the training phase we can only try to fit the model to the available data but the generalization error for new observations matters most when later using the model. This is a challenge because most ML methods allow to fit the model perfectly to data reducing the residuals to zero which is problematic for noisy datasets because the model then also captures the random

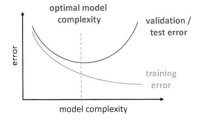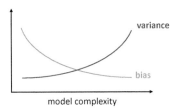

Figure 2.3: Left: training and validation/test error over model complexity. With increasing model complexity the training error will decrease; also the test/validation error will decrease until the appropriate model complexity is reached, and then as the complexity is increased even more, the test/validation will rise. Right: bias and variance over model complexity. Increasing the model complexity will lead to lower bias and higher variance.

noise which is inherently not predictable. As a consequence the prediction error on new data tends to be higher. This is called overfitting as the model fits to the training data too tightly. The opposite case is called underfitting which means that the model is not fit well enough to the training data and has not fully captured all detectable dependencies in the data.

The effects of underfitting and overfitting are visualized in Figure 2.3. The left panel highlights that we have to find the point of minimum test error which means that the model is neither underfit nor overfit. One of the variables effecting the fit is model complexity. Broadly speaking, if the model complexity is too low, we will see underfitting which means that the training and test errors are both high. In contrast, if the model complexity is too high, overfitting occurs which means that the training error is low but the actually relevant test error is high.

In simplified terms we say a model is simple if it has only few parameters and it is complex if it has many parameters. This is not the whole truth as there are other effects to consider for the tradeoff between underfitting and overfitting such as the structure of the model, the absolute size of parameter values, or the computational effort invested in fitting the model. However, in general we can say that decreasing model complexity will lead to underfitting while increasing complexity may lead to overfitting. Luckily, there are methods that allow us to detect underfitting and overfitting and which enable us to tune model complexity. For example we may use a separate validation set for estimating the generalization error.

The right panel of Figure 2.3 helps to understand how underfitting and overfitting can occur. Broadly speaking, the generalization error is the sum of two terms: model *bias* and model *variance*. The training error only accounts for the model bias and decreases for increasingly more complex models. The error because of *bias* is the error the learning algorithm makes during the modelling phase; it is seen as the difference between the predicted and the target values. Model bias can result from restrictions imposed by the model. A model with low complexity, for instance a linear model with only few parameters, has high bias. The error because of *variance* is perceived as the variability or uncertainty of the models' predictions for given data samples and increases with model complexity. The more complex the models are, the higher the probability becomes that the models are fit too tightly to data which is unwanted in the presence of noise and thus the predictions for new observations will have high uncertainty.

The relationship between them, called the *bias-variance tradeoff* is usually inverse: models with low bias tend to have a complex structure and increased variability, as they are more sensitive to fluctuations in the data. Conversely, models with high bias tend to have a simple structure and decreased variability (Hastie et al., 2001). Predictive models should ideally exhibit both low bias and variance. However, we cannot directly observe these two terms independently as we can only observe the overall error. Model variance can only be estimated by using a separate validation dataset. In general, these two sources of error cannot be simultaneously minimized.

As an example, let us discuss the following modelling scenario, which is depicted in Figure 2.4. We have a simple system with one target variable y and one independent variable x. The target y depends on x as $y = x^3 - 10\,x^2$ and is affected by noise (normal distributed noise with $\mu = 0$ and $\sigma^2 = 10$).

Without making any prior assumptions, we first employ polynomial regression (i.e., linear regression for learning the coefficients for polynomials describing the system) to predict the target variable on the basis of measured response values for $x \in \{-5, -4, \ldots, +5\}$. Since we do not know the internals of the system, we try polynomial regression for different orders. To account for potential noise, the modelling approach is repeated twenty times.

As we see in Figure 2.4 the predictions over the twenty fitted cubic models have the smallest variance and as expected fit the target function best. For higher degrees the variance of the predictions increases as a consequence of overfitting to the noisy data. For the overfit models, *extrapolation* (the application of the model outside of the data ranges used for training) leads to large errors and high uncertainty; even for *interpolation* (the application of the model to data points that lie between data points included in the training data) the overfit models may produce large errors.

We therefore have a distinction between the learning ability and the generalization ability of models and we cannot simply fit the model using all the available data. Instead, we have to estimate the generalization error. For this estimation we can subdivide the set of observations into a training set used for learning the model and a test set used for estimation of the generalization error.

In order to have an unbiased estimation of generalization ability, the test data must be completely independent and disconnected from any other aspect

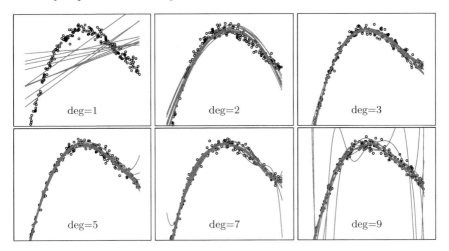

Figure 2.4: Polynomial models of different degrees fitted to random training samples with twenty data points each. Linear and quadratic functions cannot be fitted well enough (underfitting), while polynomials of higher degree have a high variance because of overfitting. The best predictions are produced for cubic polynomials as expected.

of the modelling process. That is to say, we cannot use it for any other purpose at all.

Looking at the modelling process as a whole, we additionally distinguish between optimization of model parameters during learning and the need to choose good parameters of the learning method itself. The parameters of the method are often called hyper-parameters to distinguish between the model and method parameters. Tuning the hyper-parameters requires a similar estimation of generalization ability for which we can use another dataset independent from the test set (e.g., a validation set).

These aspects must be taken into account during the experiment design phase and the available data must be appropriately partitioned using one of the partitioning schemes illustrated in Figure 2.5.

Commonly accepted supervised learning practice divides the set of available observations between:

- Training data, used for training the model parameters.

- Validation data, used to validate the learning method.

- Test data, used for assessing the generalization ability of models.

Under this formalism, models are trained using the training set, under different parameterizations of the learning method itself. The validation set is then used to compare the performances of models obtained with different hyper-parameters and select the most appropriate one. Finally, when a fully

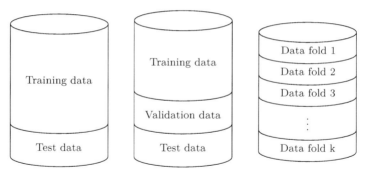

Figure 2.5: Left: data are split into training and test data. Middle: data are split into training, validation, and test data. Right: data are split into k partitions for k-fold cross-validation.

specified model is obtained, its final performance characteristics are evaluated on the test set. Usually we use the union of the training and validation sets for training the final model with optimized hyper-parameters.

2.6 Cross-validation

Cross-validation (Allen, 1974; Stone, 1974) is a widely used technique in ML for getting a detailed statistical analysis of the generalization capabilities of models created using a given modelling technique. In *k-fold cross-validation* the available data are separated in k equally sized, complementary subsets, and in each training/test cycle one data subset is chosen as test and the rest of the data as training and validation samples. This is shown in the right part of Figure 2.5.

Its main purpose is the estimation of the generalization error, a fundamental requirement for practical ML work. In cross-validation, the expectation of the generalization error is replaced by a mean over a test set $\mathcal{V}$. The procedure is unbiased as long as $\mathcal{V}$ are independent of the training sets used in the inner optimization loop for training models M.

Figure 2.6 schematically illustrates the usage of training, validation, and test data partitions, here denoted by $\mathcal{D}$, $\mathcal{V}$, and $\mathcal{T}$. The learning method takes a training data partition and accepts different combinations of hyper-parameters from hyper-parameter space $\mathcal{H}$. Each set of hyper-parameters λ is validated against $\mathcal{V}$ until λ_{best} is found.

In k-fold cross-validation, training data $\mathcal{D}$ is split into k equally-sized partitions or folds where $k-1$ folds are used for training and the final fold is used for calculating model error. The mean error in this case is given by the

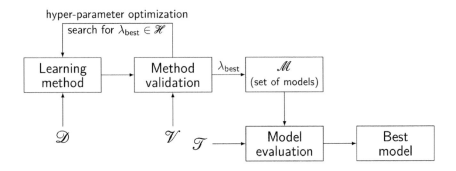

Figure 2.6: Supervised learning modelling flow with training, test, and validation data.

average over the k trials. Cross-validation therefore returns an estimate for the generalization error of the the parameterized modelling method.

In the following chapter we focus exclusively on SR giving three examples to demonstrate the basic concepts and the capabilities of SR. After that we will provide an introduction to evolutionary computation and especially genetic programming (GP).

2.7 Further Reading

We have discussed the basics of ML and supervised learning only briefly in this section. There are many good textbooks that provide a lot more details on these topics. One of the standard text books on this topic with detailed discussion of the statistical background is *The Elements of Statistical Learning* by Hastie et al. (2001). It discusses the mathematics for supervised learning and all of the standard algorithms. *Pattern Classification* by Duda et al. (2000) is similar but provides more details on neural networks. Another good reference is *Pattern Recognition and Machine Learning* by Bishop (2006) which has more details on graphical models and kernel methods. A more recent book with a focus on Bayesian methods is *Bayesian Reasoning and Machine Learning* by Barber (2012).

Data mining is related to ML and supervised learning, because its goal is to find surprising and useful patterns in databases. Regression and classification are examples of techniques for data mining that can uncover such patterns. There are many books on data mining which discuss regression and classification methods in more detail. One example is *Data Mining: Practical Machine Learning Tools and Techniques* by Witten and Frank (2005).

Several good books focus on one particular method specifically. In this context, *Gaussian Processes for Machine Learning* by Rasmussen and Williams (2006) can be recommended. *Learning with Kernels* by Schölkopf and Smola (2018) gives more details on support vector machines.

One approach to understanding supervised learning is through information theory. *Information Theory, Inference, and Learning Algorithms* by MacKay (2003) is one of the standard references in this area.

Finally, *system identification* is closely related to supervised learning but more focussed on simulation and control of dynamical systems. *System Identification: Theory for the User* by Ljung (1998) gives a good introduction to the theory.

3

Basics of Symbolic Regression

Symbolic regression (SR) is a particular type of regression analysis where the learned model is represented by a mathematical expression, which is found by manipulating a symbolic representation of that expression. By contrast to other forms of regression analysis, the model structure is not specified beforehand because it is identified simultaneously with the numeric parameters of the model. SR is therefore especially suited for regression tasks where a parametric model is not available. The obtained models can further be examined by domain experts to gain a better understanding about the modelled system.

3.1 Example: Identification of a Polynomial

We demonstrate the process of modelling a simple target function using GP. We want to identify a symbolic model for the data displayed in Figure 3.1.

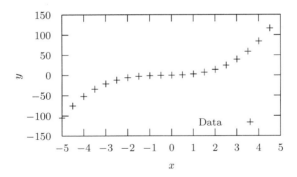

Figure 3.1: Visualization of example dataset shown in Table 3.1.

Even for simple problems, the practitioner following a purely empirical modelling approach has to take the following aspects into account: data collection and preprocessing, modelling approach, and parameterization of the algorithm.

3.1.1   Data Collection and Preprocessing

The first step is to generate some training data for this problem. We represent our data points as (x, y) tuples where $y = f(x)$. We sample 21 values of our function for x equidistantly spaced between $[-5, 5]$ and obtain the corresponding y values. The resulting dataset is organized in matrix form where each column represents one variable:

Table 3.1: Example dataset containing two variables x and y.

x	y
-5.0	-105.00
-4.5	-75.35
-4.0	-51.96
$\vdots$	$\vdots$
4.0	83.94
4.5	115.84
5.0	155.00

In supervised learning terminology, Table 3.1 represents our data, which contains input variable x and target variable y. Our task is to find a SR model M such that $\hat{y} = M(x)$ and the error between $\hat{y}$ and y is minimized. As described in Section 2.5, in a supervised learning scenario the data should be partitioned into a training and a test set.

Since our data contains a single input variable and there is no indication of data anomalies or outliers, we do not need to perform any additional preprocessing in this instance. In general, a successful approach depends on the quality of the data, presence and level of noise, and outliers or scale differences between features. These aspects need to be addressed before modelling can begin.

3.1.2   Establishing a Baseline

In this illustrative example we assume that we do not have any prior knowledge about the function we want to model. However, in real-world knowledge discovery scenarios such knowledge might be available. Therefore, before starting a potentially long and complicated modelling process, it is always good practice to try to gain more insight about the data by exploratory data analysis including data visualizations.

To get a first baseline we train a linear regression model using two-thirds of the data for training. The baseline score expressed as the R^2 coefficient of determination for the linear model is 0.838 and the linear model has the following expression:

$$\hat{y}_{LR} = 15.871\,x + 8.019 \tag{3.1}$$

3.1.3 Modeling Approach

Next, we approach the problem using SR. A number of decisions have to be made concerning the parameterization of the algorithm, such as:

- Primitive set

- Tree size and depth limits

- Population size

- Number of generations

- Selection operator

- Mutation operator and mutation probability

Although the GP practitioner can rely in most cases on a number of reasonable rules of thumb for algorithm parameterization, in general there is no configuration that performs well on every problem. Finding good parameter values for the problem at hand is a process of informed trial and error which becomes easier with experience. Practitioners have at their disposal various tools for exploring the parameter space such as grid search or other more advanced hyper-parameter tuning methods.

Primitive set: Considering the shape of the target function, starting with a basic function set $F = \{+, -, \times, \div\}$ looks promising for the first modelling attempt. With this function set we can express any kind of rational polynomial function. For terminals, we consider the set $T = \{\text{constant}, \text{variable}\}$. In this case we have only one variable, x. Constants are generated and inserted into the tree structures during population initialization and subsequently altered by mutation.

Maximum solution size: The next important decision is how large we want to allow our evolved tree expressions to become. Larger trees may lead to more accurate models at the cost of higher computational burden and with the risk of overfitting the data. For this target function, we choose relatively low limits of 15 nodes and a maximum depth of eight levels.

Population size and number of generations: The size of the evolved population needs to be sufficiently large to allow an effective global exploration of the search space. As a rule of thumb, the larger the population size, the better – assuming that sufficient selection pressure can be applied on a large population. However, since the problem we are trying to solve is relatively easy, a smaller population should be enough. We settle for a population of fifty individuals. For more complex problems poplation sizes of a few hundreds to a few thousands are typical. Larger population sizes of more than 100,000 individuals are useful only in extreme cases.

Similarly, the maximum number of generations needs to be high enough to allow the population to reach a stable state of convergence, that is when we are reasonably sure that further improvements are no longer possible. A good rule is to run the algorithm until no further improvement is achieved. We set a maximum number of 100 generations for this problem.

Genetic operators: The selection operator determines how the algorithm chooses parent individuals for reproduction. We have to specify a selection strategy (typically either fitness-proportional or ordinal) and in some cases also adjust the intensity of selection pressure applied on the population.

For example, tournament selection – one of the most popular selection operators – implements ordinal selection (individuals are ranked based on their fitness) and allows the user to tune selection intensity via the tournament size parameter. Larger tournaments increase selection pressure making it more difficult for average individuals to make it into the recombination pool. Inappropriately high selection pressure, relative to the size of the population, may lead to rapid loss of population diversity and premature convergence. This is apparent when plotting the fitness of the best individual over generations by a quick initial improvement and stagnation. Lower selection pressure usually leads to slower fitness improvements but higher fitness in the long run. In our case we employ a tournament selection scheme with a tournament size of three.

Finally, the mutation operator helps introduce new genetic material into the population, improving diversity and allowing the algorithm to explore new points in the solution space. From this perspective, the mutation intensity has an effect on the exploration vs exploitation tradeoff during the run. However, since the main operator promoting exploration is crossover, mutation should be limited to a lower intensity. A mutation probability between ten and thirty percent is usually sufficient to boost population diversity. In our experiment we employ a mutation rate of 25%.

Fitness function: We use the squared correlation coefficient ρ^2 to assign model quality as it is invariant to the scale and offset of a model's response on the training data, thus removing the burden of finding offset and scale for the model from the GP process which can be easily calculated in a postprocessing step (Keijzer, 2004).

3.1.4   Modeling Results

After the configuration phase, we run GP on the generated data and obtain a SR model that is able to perfectly explain the target. This model is illustrated along with the linear regression model in Figure 3.3. It represents a perfect fit to the target function with an ρ^2 correlation of 0.999999. The structure of the model shown in Figure 3.2 resembles a polynomial quite closely.

The formula encoded by the tree has the following expression:

$$y = ((\theta_1 x - \theta_2) + \theta_3 \, x \, \theta_4 \, x) \, \theta_5 \, x \theta_6 + \theta_7 \tag{3.2}$$

where the vector of coefficients $\boldsymbol{\theta}$ consists of values:

$$\boldsymbol{\theta} = \begin{bmatrix} \theta_1 & ... & \theta_7 \end{bmatrix} = \begin{bmatrix} 1.119 & -1.132 & 0.683 & 1.642 & 0.727 & 1.227 & 0.010 \end{bmatrix}$$

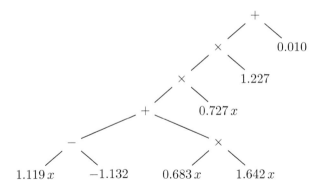

Figure 3.2: The expression tree for Equation (3.2) found by genetic programming.

We notice a degree of redundancy in the representation due to the tree constraints. These artefacts are easily removed in a subsequent postprocessing step by folding the constants θ_5 and θ_6, as well as θ_3 and θ_4. The resulting expression has a much simpler form:

$$y = x\,(\theta_1 x + xx\theta_2 + \theta_3)\,\theta_4 + \theta_5$$

with

$$\boldsymbol{\theta} = \begin{bmatrix} \theta_1 & \ldots & \theta_5 \end{bmatrix} = \begin{bmatrix} 1.119 & 1.121 & 1.132 & 0.891 & 0.010 \end{bmatrix}$$

Another trick available to the practitioner when working with SR models is to further optimize the values of the coefficients using a local search procedure such as gradient descent. After optimization, we obtain $\theta_1 = \theta_2 = \theta_3 = 1.1244$, $\theta_4 = 0.88939$ and $\theta_5 = 0$. The simplified model still has a redundant parameter (θ_4) which means that there is no unique optimum. In this example, local optimization converged to the result $1.1244 \cdot 0.88939 \approx 1$. Expanding the model leads to

$$y = \theta_1 xxx + \theta_2 xx + \theta_3 x + \theta_4$$

with

$$\boldsymbol{\theta} = \begin{bmatrix} \theta_1 & \ldots & \theta_4 \end{bmatrix} = \begin{bmatrix} 1.00 & 1.00 & 1.00 & 0.00 \end{bmatrix}$$

This indicates that the data generating function has been a simple third degree polynomial, which indeed has been used to generate the data in this simple example.

This example has shown that sometimes GP is able to discover the exact analytical form of the underlying generator function, which would not have been possible with a black-box modelling approach. It additionally highlights

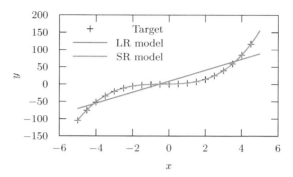

Figure 3.3: Target values y and the predictions of the linear regression and SR models.

the ability of SR models to be further simplified and processed towards better interpretability, which represents a significant advantage for engineers and practitioners in the field of system identification.

Of course, if we already know that the data can be described with a polynomial, we should fit a polynomial model directly instead of using SR. This would be much more efficient than a search over model structures with SR. In fact this example is trivially easy for GP because the generating function is short and the data is not noisy. This example is meant as a simple demonstration to show that SR is able to identify the structure of the model automatically. The example in the next section uses a noisy dataset from actual measurements and demonstrates a more realistic case where the generating function is not known.

3.2 Example: Discovery of Laws of Physics from Data

Here we use another simple example to introduce the basics of SR. The topic of this example is the *discovery of laws of physics from data*, in particular the equations of motion of falling objects. To show that ML methods are able to discover physical laws from data, de Silva et al. (2020) dropped different balls (see Table 3.2) from a bridge and captured the fall of each ball using a video camera. Each ball has been dropped twice to mitigate the inaccuracy of the data capturing process. From the videos they extracted data on the height and velocity of each ball over time.[1] While de Silva et al. (2020) were mainly concerned with finding ordinary differential equations in explicit form, we use

---

[1] Data from https://github.com/briandesilva/discovery-of-physics-from-data

Table 3.2: Weight and size of the balls.

Ball	Weight (oz)	Circumference (cm)
Baseball	5	22.25
Blue basketball	18	75
Green basketball	16	73.25
Bowling ball	81	67
Golf ball	1.62	13.8
Tennis ball	2	20.75
Wiffle ball 1	1	22.8
Wiffle ball 2	1	22.8
Yellow Wiffle ball	1.5	29
Orange Wiffle ball	1.5	29

SR to find a formula for the distance that the object fell as a function of time. We give an example for the identification of differential equations in Chapter 7.

Obviously, the balls experience different drag forces because of the differences in their surface. The Wiffle balls are light and have holes on one side which affect their aerodynamics and therefore also their falling velocity. We therefore do not use the data for the Wiffle balls.

For SR using genetic programming we only have to choose which operands, operators, and functions are allowed to occur in the model. Assuming that we know nothing about the underlying physics, we use the following rather general set of operators: $+, \times, \div, \sqrt{x}, x^2, \log(x), \exp(x)$ and allow t (time in seconds), m (weight in kg), and c (circumference in cm) as well as numeric parameters as operands.

After a few seconds GP finds

$$\text{distance} \approx \left(-0.1331\,m - 2.05\,t + 0.0016\,c\right)^2 \tag{3.3}$$

which can be expanded to

$$\text{distance} \approx 4.2026\,t^2 + 0.0177\,m^2 + 0.5456\,t\,m - 0.006564\,c\,t$$
$$- 4.2614 \times 10^{-4}\,c\,m + 2.5633 \times 10^{-6}\,c^2 \tag{3.4}$$

Ignoring drag and relativistic effects we know that

$$\text{distance} = \frac{g}{2}t^2 \tag{3.5}$$

with g approximately 9.807 on Earth.

We observe that the SR model contains the main effect with a slightly smaller factor (4.2026) which is off by 14% and has additional terms to approximate the drag effect. Interpreting the coefficients for the additional terms, we can derive that distance increases for balls with higher mass and that distance decreases with larger circumference. Both effects increase linearly

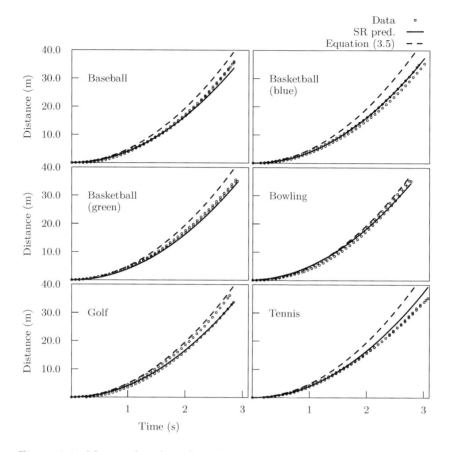

Figure 3.4: Measured and predicted distances over time for the six balls. The reference model ignoring drag is less accurate than the SR prediction.

with time. The remaining terms have relatively small coefficients and can be neglected.

Figure 3.4 shows the travelled distance over time measured for the six balls as well as the prediction from the SR model. As a reference the distance as given by the model without drag is also shown. The SR model fits the data well with a mean absolute deviation of 0.64 meter and the predictions are closer to the data than the reference model.

Interpreting Figure 3.4 we can see that the reference model fits the trajectory of the bowling ball and especially the golf ball, with its dimples to reduce drag, well. This implies that the drag forces experienced by these two balls for the 45m drop are negligible. The SR model, which can only assume an average drag coefficient, underestimates the travelled distance for those two balls. On the other hand the SR model fits the data for the baseball and the two basketballs better than the reference model. Finally, both models overestimate the velocity of the tennis ball which implies that its rough surface causes the largest drag coefficient.

Keep in mind that the SR model is a statistical model which is valid only for similar balls and conditions for which it was trained. For instance the model will certainly produce incorrect predictions if the balls are dropped from a much higher height so that they can reach terminal velocity. For instance, this model does not describe the behaviour of the Wiffle balls correctly because they experience increased drag because of their holes which is obviously not captured by the model.

In the next section we consider a different example, where we are not trying to find a formula from observational data, but instead try to approximate a transcendental function, which is difficult to calculate, using a much simpler closed-form expression.

3.3 Example: Approximation of the Gamma Function

The Gamma function is an interesting challenge for SR since for non-positive integers the target function values alternate signs between its poles, while for positive integers there are no poles but its values increase faster than an exponential function (Moscato et al., 2021).

The Gamma function Γ is a transcendental function with many applications in the area of discrete mathematics, number theory, quantum physics, or fluid dynamics. Its analytical form for complex z with positive real part is:

$$\Gamma(z) = \int_0^\infty x^{z-1} e^{-x} dt, \ \Re(z) > 0$$

The Gamma function occurs frequently in physical laws due to the prevalence of expressions of type $f(t)e^{-g(t)}$ which describe processes that decay exponentially in time or space.[2] If f is a power function and g is a linear function, the following expression can be derived via substitution:

$$\int_0^\infty t^b e^{-at} dt = \frac{\Gamma(b+1)}{a^{b+1}}$$

The fact that integration is performed along a positive real interval means that the Gamma function can be used to describe the accumulation of a time-dependent process or the total of a distribution. Because it doesn't have a closed-form solution, in numerical calculations the Gamma function is approximated through interpolation (Causley, 2022).

---

[2] Gaussian functions $a \exp\left(-\frac{(x-b)^2}{c^2}\right)$ are an important category of exponentially decaying functions.

Table 3.3: Example dataset for approximating the Gamma function.

x	y
-2.243	-1.792
-2.173	-2.581
-2.022	-22.793
⋮	⋮
4.277	8.592
4.317	9.063
4.461	11.015

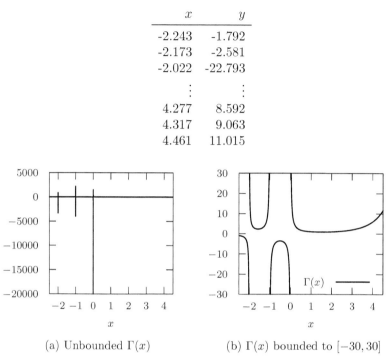

(a) Unbounded $\Gamma(x)$ (b) $\Gamma(x)$ bounded to $[-30, 30]$

Figure 3.5: Line charts of two datasets generated from the Gamma function

Here, we show that SR can be used to find a computationally efficient approximation for functions that are difficult to calculate. In a first step, we generate a dataset from a numerically accurate approximation. The Gamma function is undefined for non-positive integer values of x and grows to ∞ in the vicinity of such points, therefore we must be careful to skip these regions when generating the data for our problem. Thus, we sample data in the interval $[-2.5, 4.5]$ and reject $(x, \Gamma(x))$ tuples for which $\Gamma(x) \notin [-30, 30]$. We repeat this process to generate 500 samples that are used for training. Table 3.3 shows an example for the generated dataset. Many standard libraries provide accurate approximations for the Gamma function.

Figures 3.5a and 3.5b show the Gamma function on the domain $[-2.5, 4.5]$. This illustrates the necessity to preprocess the data in order to eliminate outliers (in this case, extreme values of the Γ function) which would have a significant detrimental effect on the learning process.

We employ SR to model the target y from the dataset we just generated (illustrated in Table 3.3). Since this target function is harder to identify than the ones in the previous examples, we use a larger population size of 1000 individuals and let it evolve over 1000 generations. To ensure enough selection pressure in the population, we use tournament selection with a group size of five individuals. We set the size and depth limit for the models to 25 nodes

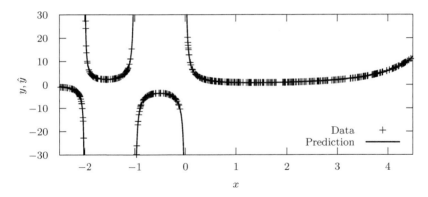

Figure 3.6: Line chart of training data and the rational approximation found with SR.

and ten levels, respectively. Besides the impact on runtime, tree size influences search dynamics by increasing the number of possible combinations of genes within the trees. This makes it possible to reach more powerful expressions, but at the same time the algorithm has to explore many more unfit models.

Since GP is a stochastic process, the usual modelling practice dictates performing multiple trials by repeating the experiment a sufficient number of times, and then collecting the best modelling results based on training accuracy from the final set of models. For this problem we repeat the experiment fifty times and select the best model. In order to increase the chances of finding a good model for this function, we employ SR with memetic optimization of coefficients (see Section 6.2).

With this experimental configuration, the algorithm is able to discover an accurate model ($R^2 > 0.999$ on the test partition) 48 times out of 50. One of the best models is

$$-0.292\,x - \frac{1.988\,x}{0.961\,x - 5.001} + 0.037 - \frac{1.017}{-2.018\,x - 4.036} + \frac{1.285}{x\,(1.29\,x + 1.29)} \tag{3.6}$$

which can be simplified to

$$-0.292\,x - \frac{2.069\,x}{x - 5.204} + 0.037 + \frac{0.504}{x + 2} + \frac{0.996}{x\,(x + 1)} \tag{3.7}$$

The estimated model $\hat{y}$ accurately predicts the Gamma function $y = \Gamma(x)$, as illustrated in Figure 3.6. What is more remarkable is that SR gives us the ability to approximate a trancendental function using an arithmetic primitive set. The algorithm finds a simple rational polynomial representation that models the data well.

3.4   Extending Symbolic Regression to Classification

If the goal at hand is not to find a regression model, but rather a classification model as described in Section 2.3, then SR has to be extended to symbolic classification. In this case, the model structure has to be extended as well as the fitness function.

3.4.1   Model Structures for Symbolic Classification

Learning Decision Trees:

The most straightforward way would be to learn models that directly calculate the class, which can be achieved by restricting the functions library to conditionals (*if/then/else* nodes) and logical functions (*and, or, ...*). The evolutionary process may then form models containing conditionals, logical functions, and variables and constants, which will lead to the production of decision trees.

Learning Discriminant Functions with Thresholds:

A way more general and flexible method is to allow the process to learn real-valued mathematical functions and then map the function evaluations to the class labels. Most naturally, GP can be used to learn models that calculate values that fit to the given class values. Thus, in that case we use SR as described in previous sections applying, for example, the mean squared error function or the coefficient of determination as quality criterion.

This implies that the class labels have to be represented as numerical values that can be used as target values. Additionally, for each model optimal threshold values have to be identified in order to be able to decide which class is estimated for a sample.

More formally, if for an n-class classification problem with ordinal class labels a model f is learned, then the thresholds represented as additional parameters $\zeta_1 \ldots \zeta_{n-1}$ have to be defined for producing the predicted class $\mathrm{class}(\boldsymbol{x})$ for input $\boldsymbol{x}$.

$$\mathrm{class}(\boldsymbol{x}) = \begin{cases} 1 & f(\boldsymbol{x},\boldsymbol{\theta}) \leq \zeta_1 \\ i & f(\boldsymbol{x},\boldsymbol{\theta}) > \zeta_{i-1} \mathrm{and} f(\boldsymbol{x},\boldsymbol{\theta}) \leq \zeta_i \\ n & f(\boldsymbol{x},\boldsymbol{\theta}) > \zeta_{n-1} \end{cases}$$

This is exemplarily shown in Figure 3.7 for four ordinal classes. Obviously, this also works for binary classification problems where a single threshold value is sufficient.

The thresholds $\boldsymbol{\zeta}$ should be set so that the resulting classifications are optimized with respect to one of the error measures defined for classification in Section 2.3 (e.g., accuracy, or sensitivity). This optimization has to be done

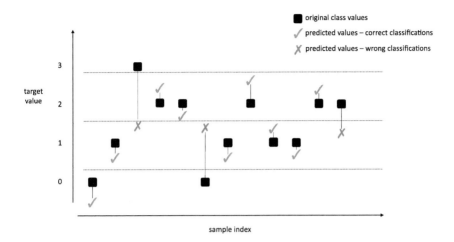

Figure 3.7: Original vs. estimated class values calculated using a regression model. The final classification is made depending on class thresholds (shown as dashed lines).

either on the training data or on separate validation data which are excluded from the training data.

Instead of using discriminative functions with thresholds, it is also common to use a nonlinear transformation to map the response of the model to the unit interval and interpret the result as the probability to belong to one of the classes. For binary classification problems, a sigmoid function can be used to map the output of the SR model $f(x)$ to the range $0 \ldots 1$, or $-1 \ldots 1$ e.g., via the logistic function

$$c(\boldsymbol{x}, \boldsymbol{\theta}) = (1 + \exp(-f(\boldsymbol{x}, \boldsymbol{\theta})))^{-1} \qquad (3.8)$$

This approach is conceptually similar to logistic regression.

For $k > 2$ classes, that cannot be put into a natural order, we can learn multiple expressions to solve the task. In the *one-vs-all* approach we learn k models that distinuish a class value from the rest. In the *one-vs-one* approach we learn $k(k-1)/2$ models that distinguish between two classes (one-vs-one).

3.4.2 Evaluation of Symbolic Classification Models

The most obvious measure for the prediction quality of symbolic classification models is the accuracy as described in Section 2.3.

An alternative to accuracy as quality function is to calculate the accuracy for each class individually. Additionally, each class accuracy can be weighted, so we use a *weighted classification accuracy* as fitness function that guides the search process.

In the case of multi-class modelling, the use of *misclassification matrices* might be beneficial, especially if misclassifications are not considered equally bad. A misclassification matrix has to define the punishment factor for each combination of classes, so that the user is enabled to define which misclassification is considered a minor problem and which misclassification is considered critical. For binary classification problems, several functions that were already described in Section 2.3 can be used.

After these four examples which demonstrate the capabilities of SR, we can now look into the details of GP in the following section, to get a better understanding of the process as well as the most relevant algorithm parameters that we have to set.

3.5 Further Reading

In this chapter the basics of SR have been introduced through modelling exemplary datasets. The book *A Field Guide to Genetic Programming* (Poli et al., 2008a) is considered another great introduction to genetic programming and SR.

The two papers *Where are we now? A large benchmark study of recent symbolic regression methods* (Orzechowski et al., 2018) and *Contemporary Symbolic Regression Methods and their Relative Performance* (La Cava et al., 2021) give a great overview of the performance of various SR methods. The latest methodological developments are presented and discussed in the yearly workshop on SR held as part of the *Genetic and Evolutionary Computation Conference* (GECCO) and organized partly by the authors of this book.

At last, to put SR into context, we recommend the book *Interpretable Machine Learning* (Molnar, 2020) that focuses on interpretable ML methods besides SR.

4

Evolutionary Computation and Genetic Programming

The previous chapters introduced the reader to supervised ML tasks such as regression and classification, and general modelling techniques to address them. The introductory examples to SR outlined the general approach without going into the details of the solution method.

In this chapter we focus on genetic programming (GP) (Koza, 1992), which is inspired by natural evolution and can be applied as a supervised learning technique for SR. GP belongs to the larger field of evolutionary computation (EC), which encompasses all optimization methods that implement aspects of natural evolution. A good understanding of the basic concepts of evolutionary computation and GP is necessary to use GP for SR with confidence. GP has many parameters which can be intimidating to set correctly for users without the necessary background knowledge. In the first few sections of this chapter we briefly summarize the basic concepts for readers who are mainly interested in using SR and GP for applications. After the basic concepts we go into the theory of EC and discuss different GP variants and operators in detail. These sections can be skipped by readers who are mainly interested in applications and can be revisited at a later time. More in-depth knowledge about GP is useful for readers who have to customize the method. This can be necessary for instance if the problem to be solved has properties that make it challenging to solve with existing GP implementations or with default settings.

In what follows, for the sake of brevity, we use the term SR to refer to the task of developing SR models using GP. SR defined as such evolves a population of mathematical expressions according to the principles of natural evolution. Expressions in the population are subject to mutation and fitness-based selection pressure. The notion of fitness is assimilated in this context to the error metric (see Section 2.2.3) characterizing the model's ability to explain the target variable. The model with the best fitness is returned by the evolutionary method as the result of the search.

Natural evolution enables populations to adapt to the environment and acquire useful traits via variation, differential reproduction, and heredity. In the context of supervised learning, this process can be seen as analogous to a learning process in which the environment and goals are artificially set according to the specific requirements of the optimization problem. This analogy first appeared in the work of Alan Turing, where it was used to outline a teaching process for machines (Turing, 1950), then later in the work of John von Neumann on self-reproducing automata (von Neumann, 1966).

The field of evolutionary computation (EC) focuses on optimization methods inspired from natural evolution. Evolutionary methods perform a guided randomized search of the solution space and are not guaranteed to reach the optimal solution even though, in practice, very good solutions are often returned. The stochastic element is required in order to have random variation, an essential component of evolution. Even though the final result of the method is not guaranteed to be optimal, we refer to it as *solution* in the following. All points of the solution space visited by the method are *solution candidates*.

Algorithm 1 shows a high-level algorithmic outline of the evolutionary process. Search strategies with the general structure given in Algorithm 1 differentiate themselves from one another by the way solution candidates are encoded (the chromosome representation) and by the concrete implementations of subordinate procedures such as recombination and selection. The recombination step may be performed by multiple different recombination operators applied with different probabilities. In evolutionary methods solution candidates are also called *individuals*.

Algorithm 1 General workflow of evolutionary methods

1: **initialization**: initialize population with random individuals
2: **fitness evaluation:** assign fitness values to the initial individuals
3: **while** stopping criteria not satisfied **do**
4:      **parent selection:** select parent individuals for recombination based on fitness
5:      **recombination:** generate new children from the selected parents
6:      **fitness evaluation:** calculate objective values for each child
7:      **replacement:** fill next population with children and individuals from the current population based on their fitness
8: **return** best individual from the population

The most popular evolutionary methods include evolutionary programming (Fogel and Fogel, 1996), evolution strategies (Rechenberg, 1971; Schwefel, 1977), genetic algorithms (Holland, 1975), differential evolution (Storn and Price, 1997) and GP (Koza, 1992).

GP distinguishes itself from other evolutionary methods by evolving computer programs that incorporate problem-solving behaviour. The generality of this approach makes GP a domain-independent optimization method with many application domains where simple programs or functional expressions have to be found. Examples are optimization of dispatching rules for production or logistics, evolving expressions for feature engineering, evolving ML pipelines for automated ML (AutoML), and of course SR.

Figure 4.1 shows the general outline of the GP evolutionary process and describes the interplay between its subordinated procedures.

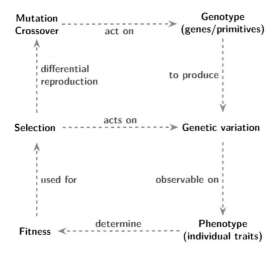

Figure 4.1: GP structure and operator relationships.

The GP operators *crossover* and *mutation* produce new child individuals from selected parent individuals randomly:

- Crossover creates a child from parents by randomly combining parts of their encoded representation.

- Mutation creates a new child by performing a random change in the encoded representation of the parent

At each generation, individuals are evaluated, selected for reproduction, then crossed over and mutated in order to fill the offspring population. In the more general evolutionary loop described by Algorithm 1 the *recombination* step encapsulates all the logic for generating a new offspring from the current population, which may include crossover, mutation, or other operators. Many GP variants also include a neutral copy operation where children are exact copies of parents. The effect is similar to keeping some of the individuals from the current population when generating the new population.

The iterative cycle of GP is described graphically in Figure 4.2. The cycle begins after the initialization of the initial population and continues until a stopping criteria is reached.

GP has been popularized by Koza (1992) through a series of books, after earlier work by Cramer (1985) and Dickmanns et al. (1987) which described similar methods. In Koza-style GP computer programs are encoded as *symbolic expression trees* where internal nodes represent *functions* and leaf nodes represent their operands: variables and constants described together as *terminals*. Formally, the set of allowed internal symbols represents the function set $\mathcal{F}$ of the algorithm, while the set of allowed terminal symbols represents the terminal set $\mathcal{T}$. We call the set $\mathcal{P} = \mathcal{F} \cup \mathcal{T}$ the primitive set of the

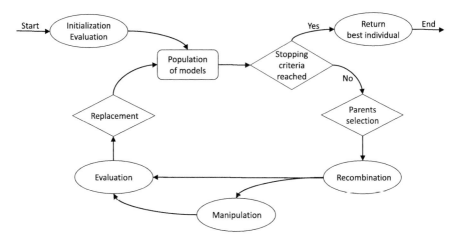

Figure 4.2: GP cycle.

algorithm. Recombination operators directly work on the tree representation. Correspondingly, we refer to this style of GP as tree-based GP.

Even though many different GP variants with other encoding schemes have been described since its inception, tree-based GP is still the most prevalent in academic publications. Correspondingly, we also mainly focus on tree-based GP in the following sections. At the end of this chapter we briefly discuss other alternatives of which linear and graph-based encodings are most important.

Figure 4.3 shows an expression tree for the formula $\cos x + i \sin x$. The ability to encode different kinds of mathematical expressions depends on the allowed set of operators for the tree's internal nodes. In general, we want to enable the GP system to evolve solution candidates that can solve the given task by including all the necessary operators and terminals in the primitive set. At the same time we try to keep the primitive set as small as possible because adding primitives increases the search space and may make the problem harder to solve.

The size and shape of the trees generated and manipulated by GP depends on properties of the primitive set $\mathcal{P}$ such as the required number of arguments of functions and operators. The properties of the encoding influence the dynamics of the GP evolutionary process and play an important role in the performance of the algorithm.

The primitive set $\mathcal{P}$, together with the evolutionary operators, determines the algorithm's ability to explore the search space. From this perspective, the primitive set is required to have two important properties:

1. *Closure.* Any combination of instructions must result in a valid program, that is any combination of sub-expressions must result in a valid expression. This property is necessary due to the way GP generates new solution candidates via recombination.

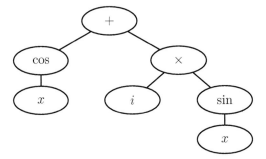

Figure 4.3: Example symbolic expression tree encoding the mathematical formula $\cos x + i \sin x$.

2. *Sufficiency.* The primitive set $\mathscr{P}$ has to be comprehensive enough to allow solutions to be expressed by the algorithm. For example, the set of Boolean functions $\{\wedge, \vee, \neg\}$ is sufficient for any problem in the Boolean domain.

   Goldberg (1989) suggests that the user should select the smallest alphabet that permits a natural expression of the problem. A larger, redundant primitive set may artificially increase the size of the search space and hinder the algorithm's ability to produce good solutions.

4.1 General Concepts

Terminology and concepts from the biological domain are often used to reason about the dynamics and performance of evolutionary algorithms (EA) including GP.

Search behaviour is often discussed in terms of population properties such as *diversity* and *evolvability*, properties of the representation such as *buffering* or *redundancy*, or properties of the search space such as *neutrality*.

These properties are usually not directly measurable and controllable, but depend indirectly on algorithm parameterization and problem characteristics. In what follows we provide a summary of the biological concepts relevant to GP performance and discuss their implications from an algorithmic perspective. An understanding of the implications is relevant for the correct configuration of the algorithm for SR.

4.1.1 Genotype, Phenotype, and Semantics

In line with biological terminology, the *genotype* refers to an individual's actual genetic structure. The structure of the genotype is determined by the chosen solution encoding and the data structures used to represent it. Typical data structures for encoding genotypes are binary vectors, real vectors, permutations or expression trees.

The *phenotype* refers to traits or characteristics of the individual that are observable or can be made visible by some technical procedure. Within GP, phenotypic traits become observable through the evaluation of the genotype structure within the context of the problem to be solved. Some encodings such as S-expressions have a genotype representation (expression tree) that is very close to the phenotypic representation (the mathematical expression), however other representations such as the BNF grammar used by grammatical evolution require an explicit translation step from linear genotypes to phenotypes. It is worth mentioning that the terms genotype and phenotype are not used consistently in GP literature. In the following sections we will use the term genotype to refer to the trees in tree-based GP.

The interaction between genetic operators acting at the genotype and phenotype level such as crossover, mutation, and selection represents the core part of an evolutionary metaheuristic. This follows from natural selection, as phenotypic variation due to the underlying heritable genetic variation is a fundamental prerequisite for evolution. Consequently, the way genotypes are mapped to phenotypes has a great influence on the dynamics of the search and on the quality of its outcome.

Adding behaviour to the phenotype as an observable characteristic (Hu et al., 2020) represents a useful extension to the definition of the phenotype and allows a better characterization of evolutionary dynamics determined by the genotype-phenotype map. In GP, phenotypic behaviour is an expression of the semantics (given e.g., by a certain configuration of operators/functions within the genoytpe structure) combined with certain inputs. So while the semantics are inherent to a GP program, its observable behaviour is quantified by its response on the training data.

In SR, observable behaviour might include model response characteristics such as nonlinearity, poles, monotonicity, boundedness, etc., along with other traits such as the size of the program, the number of different variables used in the expression, the complexity of the expression, or the corresponding functional mapping. All these characteristics are typically aggregated together into a scalar fitness value which is used by selection to decide which individuals will pass on their traits to the next generation.

It is important to note that the genotype-phenotype mapping is *non-injective*, as many different genotypes can be decoded to the same phenotype and thus be semantically indistinguishable from each other.

4.1.2 Diversity and Evolvability

Similar to population genetics, diversity in GP is understood as the amount of different individual characteristics (both structural and semantic) in the population. Selection acts towards reducing diversity as a consequence of the fact that only a subset of the population (the most fit) gets to participate in reproduction, leading to an information loss and a gradually a less-diverse recombination pool. It was shown that loss of diversity is entirely due to not-sampled individuals (Xie, 2009). McPhee and Hopper (1999) find that "progress in evolution depends fundamentally on the existence of variation in the population". Moreover, although selection acts on phenotypes, phenotypic variation is the product of both genetic and epigenetic processes.

There is frequent reference in GP literature to the importance of population diversity, as well as a plethora of proposed algorithmic enhancements towards this purpose. The population's potential to produce fitter programs through genetic operators (Burke et al., 2004; Črepinšek et al., 2013) is known as *evolvability* (Hu and Banzhaf, 2016a,b) and plays an important role in search performance. If the population is not sufficiently diverse, genetic operators such as crossover and mutation become less effective in producing novel solution candidates from the available gene pool, reducing the changes for successful adaptation. In many cases, algorithm performance can be improved by more efficiently exploiting existing population diversity during the evolutionary run.

The prevailing opinion is that diversity promotes adaptation and helps avoid evolutionary stagnation and premature convergence. Without a sufficiently-diverse pool of genetic material at their disposal, genetic operators such as crossover and mutation can become unable to produce fitness improvement. If this situation occurs before a good solution is found, the algorithm is said to have prematurely converged. One of the earliest accounts (Eshelman and Schaffer, 1991) defines *premature convergence* as a loss of population diversity before optimal or at least satisfactory solutions have been found and describe it as a serious failure mode for evolutionary methods.

The rate of diversity loss through the action of the selection operator can to some extent be controlled via the intensity of selection or selective pressure. For example, the tournament size in the case of a tournament selection scheme controls the intensity of selection, with larger tournaments leading to more intense competition between individuals and fewer unique individuals in the recombination pool. The tournament group size must be adjusted together with the population size.

The topic of whether promoting diversity at the structural (genotype) or semantic level is better for the search process remains open for debate. Structural measures use tree distances or metrics which are usually computationally expensive. Semantic or behavioural measures use information about an individual's fitness or its output for the training data and are usually easier to compute since this information is already available during evaluation.

4.1.3 Buffering, Redundancy, and Neutrality

It was noticed from the early days of population genetics that the dependency between genotypes and phenotypes is a complex one and that *epistasis* plays an important role in the the outcome of genotypic interactions (Wright, 1967). Epistasis describes the situation where the expression of a gene and its subsequent selective value depends on other genes in the genome. To better understand this dependency, Wright introduced the concept of a *fitness landscape*, defined as an imaginary surface in genotype space where neighboring points represent genotypes that differ from each other by a single mutation and the height of the surface is given by the mean population fitness of a genotype.

Within a given fitness landscape, natural selection causes the population to move towards the nearest peak, that is, to climb the fitness surface towards the nearest local optima. In this context, Wright correctly conjectured the existence of a mechanism by which the population can escape these local optima in order to be able to continually adapt. The capacity to overcome local optima represents a desirable property not only in biology but also in optimization algorithms.

Since genes are indirectly selected on the basis of observable phenotypic traits, it follows that the relationship between genotypes and phenotypes – the so-called genotype→phenotype (G→P) map – plays a role in the population's ability to cross valleys in the fitness landscape. Indeed, the non-injectivity of the G→P map implies the existence of sets of neighboring points in the fitness landscape that express the same phenotype (i.e., the same set of observable traits, quantified by fitness). A set of such points represents a *neutral network* in the fitness landscape. It was later shown that neutrality is an important prerequisite for *evolvability* – the ability to evolve past the local optima (Wagner and Altenberg, 1996).

Here, we use the term robustness to refer to a system's capacity to maintain its function against perturbations. Genotypes are robust if they maintain their phenotype under mutation, therefore robustness is an important prerequisite for neutrality and evolvability. The main underlying mechanism for robustness is genetic buffering, defined as the accumulation of hidden genetic variation that is only expressed when the genetic background changes. In practical terms, as the genotype evolves it will accumulate some genes via mutation (or crossover) that will not be expressed (i.e., they have no observable phenotypic effect) but will act as buffers against perturbation.

4.2   Population Initialization

As seen in Algorithm 1 the first step in an evolutionary algorithm is random initialization. In GP, random initialization of tree individuals has to take into account encoding specific aspects such as the *size* and *depth* of the produced trees, as well as the overall distribution of function and terminal symbols in the initial population. We use the following terminology to describe properties of trees:

- The *root node* represents the topmost node of a tree.

- The *node depth* is equal to the number of edges on the longest path from the root node to this specific node.

- The *tree depth* is the maximum depth of all tree nodes.

- The *tree size* is equal to the total number of nodes (including the root node) contained in the tree.

- *Node arity* represents the number of child nodes accepted by a function node.

Generally, initialization should aim to generate high diversity because it is the main source of genetic material in the population. Operators in later stages mainly recombine parts from genotypes. Many different methods for initialization have been described in the literature with the aim to improve GP performance. In the next section we describe some of them in more detail. However, while it is certainly possible to make a severe mistake in the initialization of GP populations (e.g., by generating only very small trees or many duplicates), we found that the choice of initialization operator usually only has a small effect on GP performance and can be countered by adjusting other parameters (e.g., population size).

We recommend checking the empirical distributions of variables, numerical parameters, and function symbols in the initial population. Each variable should occur with similar frequency and all allowed functions or operators should occur with similar frequencies. Additionally, we may check the frequency of numeric parameters relative to the frequency of variables as well as the empirical distribution of parameter values. Numeric parameters should span a larger range so that they can be combined by GP to produce a large set of parameter values. It may hamper performance if the initialization method produces only a small set of parameter values.

4.2.1   Operators

Early initialization methods proposed by Koza (1992) generate random trees from the available primitive set up to a maximum user-specified depth limit, sharing a top-down approach where trees are "grown" starting from the root node by adding randomly chosen child nodes. While the first method called `grow` may generate trees of any shape and size, the second method called `full` generates complete trees by restricting leaf nodes to the last level of the tree. A combination of the two methods called `ramped half-and-half` (half of the trees initialized with the `grow` method, the other half with `full`) was also proposed by Koza to improve the diversity of the initial population.

Both methods are sensitive to the properties of the primitive set (function arities, number of functions, number of terminals), although this aspect can be controlled by probabilities for choosing different types of nodes during tree creation. A typical solution is to choose between function and terminal nodes in the `grow` method with probability $p = \frac{|\mathscr{T}|}{|\mathscr{T}|+|\mathscr{F}|}$ where $\mathscr{F}$ and $\mathscr{T}$ are the function and terminal sets. A large population size was also suggested to ensure a more uniform distribution of tree shapes and sizes and avoid sampling artifacts. Alternatively, Böhm and Geyer-Schulz (1996) used precomputed information to uniformly sample all possible trees of a given size. Their algorithm was based on previous work by Alonso and Schott (1995) on the random generation of combinatorial objects.

Langdon (2000) observed that GP search spaces are partitioned by the ridge in the number of programs versus their size and depth and proposed a ramped uniform random initialization. His tree initialization algorithm also derived from Alonso and Schott (1995) generates a uniform range of tree sizes. According to Langdon, the ramped uniform method represents an improvement over Koza's ramped half-and-half as it produces trees with shapes near the ridge in the search space.

Luke (2000) identified several weaknesses in the tree creation methods introduced by Koza: no control over the probabilities that certain function symbols are included (the methods sample the available functions uniformly), no control over the generated tree shape (in the case of the `grow` method), and no possibility to create trees with a fixed or average tree size or depth. He proposed two new probabilistic tree creation methods named `PTC1` and `PTC2` that allow the user to specify an expected tree size and a probability distribution for the desired function nodes.

In a follow-up paper Luke and Panait (2001) found that `PTC1` and `PTC2` do not offer any significant advantage in terms of solution quality as uniformity does not have a big influence on improving fitness. However, the possibility to hand-tune function probabilities and expected tree sizes during initialization is likely to have a positive impact on algorithm dynamics and on the sizes of the resulting solutions and their interpretability.

The balanced tree creator (BTC) (Burlacu et al., 2020) is another method that can generate trees of any given size and allows control over the distribution of function nodes. The algorithm iteratively expands a randomly chosen root node in breadth-wise fashion until the target length is reached, keeping track of the remaining number of nodes to be filled. Due to the breadth-wise expansion, the resulting trees tend to be balanced; to increase the shape variability, an *irregularity bias* parameter is provided to control the probability that a leaf node is sampled instead of a function node within the expansion loop.

Depending on the available prior knowledge about the problem to be solved, a more fine-grained control over the initialization process might sometimes be desired. Although initialization algorithms are stochastic, the distribution of tree sizes can be modeled as a *branching process* (Grimmett and Stirzaker, 2020). This makes it easy to control the outcome of initialization by tuning the primitive set, the sampling frequencies of symbols, or the maximum allowed depth.

4.3 Fitness Calculation

Fitness in SR is mainly determined by the agreement (goodness-of-fit) of the model output and the target values and can be measured using one of the metrics described in Section 2.2.3. For non-trivial problems, fitness calculation is the most computationally-intensive part in a GP system, due to the necessity of evaluating all generated individuals on the training data.

Tree-encoded individuals are evaluated over the training data by executing the instructions contained in their genotype for each training sample. For example, calculating an individual's fitness over a dataset consisting of 10,000 samples usually involves executing the same program 10,000 times, each time with different input values given by the corresponding rows.

In practice, fitness calculation is typically parallelized on today's multicore machines. Since individuals are evaluated independendly from one another, fitness calculation is considered to be an "embarassingly parallel" process. Improving the evaluation speed of GP individuals, potentially by using data-parallel hardware such as GPUs (computer graphic cards), is a topic of ongoing research (Baeta et al., 2021).

4.4 Parent Selection

Selection in GP refers to the process of selecting the parent individuals which will produce the next generation of solution candidates through recombination. In accordance to the principles of natural selection, the fittest individuals in the population should take part in this process with higher probability.

4.4.1 Operators

The selection operator is one of the most important GP operators and it can be worthwhile to experiment with a few different options. Many different operators have been proposed and studied over the last decades (Blickle and Thiele, 1996).

In *fitness-proportional selection*, the selection probability for an individual depends on its relative fitness compared to the overall fitness of the population. For a population of N individuals, the relative fitness and selection probability of individual i with fitness value Fit_i is given as

$$p_i = \frac{\text{Fit}_i}{\sum_{j=1}^{N} \text{Fit}_j} \tag{4.1}$$

whereby it is assumed that all fitness values Fit_j are positive.

Fitness-proportional selection does not offer an easy way to control selection pressure. Selection pressure is a result of the distribution of fitness values and varies over the GP run. A drawback is that if differences between highly fit and lower fit individuals are large, the highly fit individuals will dominate the selection process, decreasing population diversity quickly.

Ordinal selection methods rank individuals in the order of increasing fitness and use this information to decide which individuals get selected. Since only the order is considered, the approach is not sensitive to the absolute fitness values. Popular ordinal selection methods include *ranking selection* and *tournament selection*.

Ranking selection assigns selection probabilities as a function of rank. Linear (Back, 1994), exponential (Blickle and Thiele, 1996), or polynomial (Hingee and Hutter, 2008) functions can be used to map ranks to selection probabilities.

Tournament selection (Miller and Goldberg, 1995) forms tournaments between individuals by sampling k individuals from the population. The individual with the best fitness (the tournament winner) is then selected for reproduction.

Truncation selection (Mühlenbein and Schlierkamp-Voosen, 1993) is an approach inspired from selective breeding where only the top percentage of the population is reproduced.

Ordinal selection schemes such as tournament selection apply constant selection pressure on the population (since they only consider order relationships and not actual fitness). Furthermore, tournament selection provides a tuning mechanism for selection pressure via the tournament group size, where larger tournaments apply higher selection pressure on the population.

Lexicase selection (Spector, 2012; Helmuth and Spector, 2013) considers the performance of solution candidates on individual test cases, in order to identify those that perform especially well in certain regions of the problem. The original lexicase selection mechanism is particularly well-suited for discrete error spaces but does not perform well on continuous-valued problems. ϵ-lexicase (La Cava et al., 2016) is adapted for regression problems by modulating the pass condition on test cases using a threshold parameter ϵ.

4.4.2   Selection Pressure

We define *selection pressure* as the bias of the selection scheme towards more fit individuals. A selection scheme that is highly likely to select a more fit individual over a less fit one is said to exert high selection pressure on the population, making for a more intense competition for survival.

The appropriate value of selection pressure depends in practice on the characteristics of the problem and other parameters of the algorithm such as the population size. Too high selection pressure might lead to a few fit individuals dominating the whole population and *premature convergence* of the algorithm due to loss of diversity, while too low selection pressure might not provide sufficient drive for the search to succeed.

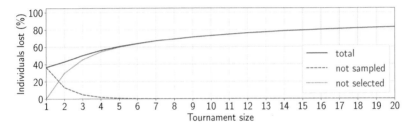

Figure 4.4: Selection pressure increase with tournament size, illustrated as the amount of programs lost.

Visualizing the best and average fitness values within the population over generations can be helpful to chose and parameterize a selection operator. When we use a selection operator which exerts high selection pressure we will see a rapid improvement of fitness followed by stagnation as the entire population will likely contain copies of the same individual. In the other case, low selection pressure (e.g., with random selection in the extreme case) will cause a slower improvement in fitness values and slower diversity loss. This may allow us to find better solutions as the process has more opportunities to recombine genotypes. We discuss how to analyze and track diversity in a GP run in Section 4.8.

When looking at tournament selection – the most popular selection scheme – Xie and Zhang (2011) observed accelerated loss of diversity with increasing group size, to the point where the percentage of programs lost due to never getting sampled or never winning a tournament quickly exceeded 50% of the population. This is illustrated in Figure 4.4 which shows that even small group sizes can lead to more than 40% of the individuals being lost.

Note that for ordinal selection schemes such as tournament selection, the rate of diversity loss remains constant over the generations. This effect, combined with the fact that genetic operators are most of the time deleterious, explains why population diversity is so hard to come by after a few generations, and why preserving diversity is important in evolutionary algorithms.

Therefore, an appropriate amount of selection pressure must be used (e.g., by setting the tournament size) for each problem and algorithm configuration.

4.5 Bloat and Introns

Program size has been shown to offer an evolutionary advantage in GP (Dignum and Poli, 2007), as larger individuals are less vulnerable to deleterious changes, more likely to achieve a better fitness and can more easily preserve their fitness. The positive correlation between size and fitness creates a tendency for individuals to grow as the generational search progresses under selection pressure, until a maximum size limit is reached.

Empirical evidence shows that in the course of evolution, as the individuals in the population become larger, recombination events such as crossover and mutation generate diminishing returns in terms of fitness, to the point where the genotypic changes they produce have a very slight or even neutral effect. Langdon (2021) observed that floating-point operations performed for internal nodes of GP trees incur a small loss of information on average. As a consequence changes at the lower levels of deep trees frequently have no effect on the output.

The phenomenon of uncontrolled growth of program sizes without corresponding improvements in fitness is called *bloat* in GP literature. The parts of GP programs that have no impact on semantics are called *introns*. Expressions such as $f(x)*0$ or $x-x$ have been given as examples for introns in GP literature. However, these cases are atypical for SR. We define *introns* as subtrees that can be deleted without changing the behaviour of the overall expression. SR solution candidates frequently contain subtrees with very small impact for the overall output. As per the definition these are not introns but make the solution candidates bloated. Subtrees with small impact on the output can be removed in a postprocessing step (pruning) to simplify the solution with a minimal loss in accuracy. We discuss model inspection and postprocessing in Chapter 5.

Bloat – the unwarranted increase in the complexity of the representation – is considered detrimental to the search process for a number of reasons:

- Genetic operators acting on bloated individuals may be less effective due to the large amount of redundancy.

- The algorithm may take longer to converge.

- Evaluating larger models requires more computational effort and can slow down the algorithm.

- Interpretability and explainability are compromised.

In this context it is important to distinguish between bloat and overfitting. Both are unwanted effects that may occur as expressions grow too large. It is easy to distinguish the two effects by looking at the improvement of fitness with larger expressions. We have bloat if fitness stagnates with growing expression sizes, while fitness still improves in the case of overfitting. In contrast, overfitting occurs when the error for a hold-out set increases with growing expression sizes. Bloat and overfitting can occur independently – we can indeed have bloat without overfitting and overfitting without bloat.

It should be noted however, that not all complexity is bad, and a bit of bloat can help to reach better and more accurate solutions. The redundancy associated with bloat can serve as a reservoir of genetic variation inside the genotype, consisting of dormant genes that can be activated by recombination operators and produce adaptive effects when placed in the right context.

Bloat has been an active research topic in the GP field for many years, leading to a number of theories for its occurrence such as replication accuracy (McPhee and Miller, 1995), removal bias (Soule and Foster, 1998), fitness distribution (Langdon et al., 1999), crossover bias and program size distribution (Dignum and Poli, 2007; Poli et al., 2008b; Silva et al., 2012).

The most widely known and used technique to control bloat is the *parsimony pressure* method, which incorporates size into fitness in the form of various heuristics e.g., by applying a fitness penalty proportional to program size (Zhang and Mühlenbein, 1995). A simple mechanism is to reject offspring that exceed a given maximum size and to recreate new offspring via selection and recombination steps until size constraints are fulfilled.

Simplification (or *pruning* of trees) represents another technique that has been employed with varying degrees of success (also see Section 5.3.2). The idea is to eliminate redundancy via simplification rules such as algebraic simplification (Wong and Zhang, 2006), hash-based simplification (Zhang et al., 2006), numerical simplification (Kinzett et al., 2009) or based on statistical tests (Rockett, 2020). Simplification reduces diversity and thus must be used carefully.

While all methods have their merits and can reduce bloat and improve model quality on certain problems, there is no general approach that automatically evolves individuals of the necessary size and frees us from configuring size targets manually. Instead we have to parameterize size limits, heuristic methods for parsimony pressure, or simplification approaches for the problem at hand. This can be difficult as the necessary size to solve the problem is usually not known beforehand.

4.6 Crossover and Mutation

Crossover and mutation are genotypic operators (i.e., they manipulate the tree structure) responsible for producing new child individuals (see Figure 4.2).

The crossover operator takes two parent individuals and produces two children by combining parts from the first and the second parent. For a fixed-length encoding, single-point crossover uses a single random cutting point in both genotypes and crosses the two parts from both parents to produce two children. The equivalent operation for tree-based GP is subtree crossover. It randomly chooses a cutting point in the first and in the second parent and exchanges a subtree from the first parent with a subtree from the second parent as illustrated in Figure 4.5. The main purpose of crossover is to enable the algorithm to explore new points in the search space by combining parts of existing solution candidates.

Since the child inherits the root of its tree structure from the first parent, this parent is also called the *root* parent and the second parent is also called the *non-*

root (or donor) parent. Inside a tree individual, a subtree's usefulness depends largely on the surrounding context. Therefore, the subtree swapping operation performed by subtree crossover can have a big impact on the semantics of the child individual.

Context-aware crossover operators (Majeed and Ryan, 2006) try to prevent bad subtree swaps by considering the context, in order to increase the chance of producing a fit offspring. The idea is to select one subtree at random from the second parent and generate a set of potential children corresponding to each possible swap location in the first parent. Then, the child with the best fitness is kept as the outcome of the crossover operation.

Semantically-aware crossover (Nguyen et al., 2011) works under the assumption that the exchange of subtrees between crossover parents is more likely to be beneficial if the subtrees are not semantically identical, but also not too different. A measure of semantic similarity between subtrees (given a set of input data) is defined in order to guide this operator. Two subtrees are semantically similar if this measure lies within a positive interval $[\alpha, \beta]$.

The mutation operator takes a copy of a parent and performs a random change or alteration of its genotypic structure, such as:

- Applying a perturbation to a numerical node. Because it affects a single node, this is called a *single-point* mutation. Alternatively, *multi-point* mutation can simultaneously perturb multiple or all numerical nodes in the tree.

- Changing a node's symbol with a compatible one (for example, with a symbol of the same arity for a function node, or a different variable symbol for a leaf node).

- Replacing a subtree with a new, randomly generated subtree. We can distinguish here between pruning operations such as replacing a subtree with a single leaf node, or expansion operators such as replacing a leaf node with a generated subtree.

In EC, effects of mutation should be small compared to effects of crossover. Mutation can act as a constant source of diversity while crossover combines already existing genetic material.

Both crossover and mutation operations may implement additional logic to prevent generating offspring that violate the tree depth and size limits set by the user.

4.7   Power of the Hypothesis Space

The hypothesis space represents a infeasibly large space of possible combinations of symbols from which solution candidates are assembled by the GP algorithm.

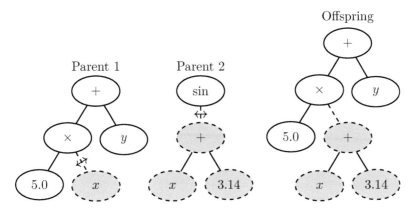

Figure 4.5: The crossover operation, where ↔ represents the crossover point.

The most important property of such a space is that it must contain a solution to the problem and that this expression must be reachable by the algorithm through the successive application of its genetic operators, starting from a random initialization of the population.

In this context, the *power* of the hypothesis space can be intuitively seen as the ability to quickly and effectively reach the correct combination of symbols from the random starting points. Here, prior knowledge can be integrated into the process by configuring the primitive set and genetic operators to restrict the hypothesis space and increase the efficacy of the search. GP is a very generic solution method and especially the primitive set has to be customized for individual problems. The ability to effect the power of the hypothesis space via expert knowledge can be considered as an advantage of GP compared to other learning methods. Customization of the primitive set offers a straightforward way to customize the search process for a particular application domain.

Evolutionary methods including GP can be scaled up by increasing the population size and the generational limit in order to tackle difficult problems. Fitness evaluation can be parallelized easily and even multiple populations can be evolved on separate computers with minimal information exchange required between the parallel processes. At the same time, increasing size limits allows GP to develop more complex solutions.

If the hypothesis space is powerful enough, problems of any complexity can theoretically be solved with a sufficiently large population, running for sufficiently many generations. In practice, however, we are limited by time and resource constraints, which means that there is a limit to the problem complexity for which we can hope to find an acceptable solution.

The choice of algorithm parameters plays an important role in the modelling process. Inappropriate configurations can end up wasting computational resources (for example, when the algorithm runs for 1000 generations, but is already converged after 100 generations), producing overfit models (when the

Table 4.1: Algorithmic configuration

Population size	1000
Generations	500
Selection mechanism	Tournament selection (group size 5)
Fitness measure	NMSE

solution candidates are allowed to become too large and complex), or simply not finding a good model at all.

To avoid these problems it is necessary to configure the algorithm's primitive set and its genetic operators such that the hypothesis space is sufficient for the problem, but not overly large.

For example, if the data can be optimally modeled by the rational function (Pagie and Hogeweg, 1997):

$$y = f(\boldsymbol{x}) = \frac{1}{1 + x_1^{-4}} + \frac{1}{1 + x_2^{-4}}, \quad x_1, x_2 \neq 0 \tag{4.2}$$

then the hypothesis space defined by a primitive set $(+, -, \times)$ for polynomials does not have enough expressive power to enable the algorithm to find an acceptable solution. To illustrate this problem we design an experiment using the configuration described in Table 4.1 and use a dataset where we sample $\boldsymbol{x}$ from the domain $[-5 \ldots 5]^2$ without zeros.

Using these settings, the best model out of fifty repetitions achieves a training average relative error 30.4% and a test average relative error of 32.1%, and has the following expression:

$$f_{\text{Poly}}(\boldsymbol{x}) = x_1^4\,\theta_1 + x_1^2\,\theta_2 + x_2^4\,\theta_3 + x_2^2\,\theta_4 + \theta_5 \tag{4.3}$$

with coefficient vector

$$\boldsymbol{\theta} = \begin{bmatrix} -0.003 & 0.104 & -0.00332 & 0.1333 & 0.677 \end{bmatrix} \tag{4.4}$$

It is clear in this case that the algorithm does not have the power to evolve an acceptable solution for this problem, due to an insufficiently powerful hypothesis space.

When the primitive set is augmented with the $\div$ symbol, the algorithm is able to find a much better solution for the problem. The best run out of fifty repetitions achieves an average relative error of 0.8% on both training and test set. In this case, the following expression is discovered:

$$f_{\text{Rat}}(\boldsymbol{x}) = x_2^2\,\theta_1 + \frac{1}{x_1^4\,\theta_2 + \theta_3} + \frac{1}{x_2^2\,\theta_4 + \theta_5} + \theta_6 \tag{4.5}$$

with coefficients

$$\boldsymbol{\theta} = \begin{bmatrix} -0.009 & -1.0 & -1.0 & -0.624 & -0.740 & 2.255 \end{bmatrix} \tag{4.6}$$

When the hypothesis space is overspecified by additionally including the *exp* and *log* symbols in the function set, the algorithm achieves a marginally better model quality at the cost of a significant increase in complexity through nested *exp* and *log* functions. The best model out of fifty repetitions achieves a training and test average relative error of 0.7% and has the following structure:

$$f_{\text{SR}}(\boldsymbol{x}) = \log\{\log[\exp(x_1^2\,\theta_1)\,\theta_2 + \exp(x_2^2\,\theta_3)\,\theta_4 + \theta_5]\}\,\theta_6 + \theta_7 \qquad (4.7)$$

Due to the nesting of exponential and logarithmic functions, the formula has to be scaled by large coefficient values and the evaluation becomes numerically unstable which has the effect that using the rounded coefficient values given below instead of the values with full precision leads to larger prediction errors.

$$\boldsymbol{\theta} = \begin{bmatrix} -0.786 & 33.278 & -0.762 & 33.107 & 5.6 \cdot 10^5 & -2.4 \cdot 10^5 & 6.2 \cdot 10^5 \end{bmatrix}$$
$$(4.8)$$

This example nicely demonstrates that it is worthwhile to pay attention to the appropriate definition of the hypothesis space via the primitive set. It is a great example for the bias-variance-tradeoff in SR (see Section 2.5) where the power of the $\mathscr{P}$ affects model complexity and therefore the variance of the model. While an underspecified hypothesis space is usually easy to detect due to the low fitness achieved, an overspecified hypothesis space may have more subtle effects on the evolutionary search process, such as delayed convergence, increased risk of overfitting, or increased solution complexity, with a detrimental effect on model interpretability.

As illustrated by the examples, the three different hypothesis spaces generated different model structures. In the first case, removing a critical symbol in the primitive set (the ÷ symbol) took away the algorithm's ability to find a good approximation of the target function. In the third case, adding redundant symbols to the primitive set (the *exp* and *log* symbols) caused the algorithm to generate an overly-complex model. In general, however, we do not know which symbols are required in the primitive set for a given problem. Therefore, narrowing down the hypothesis space to an optimal size is a process involving some trial and error, supported by empirical validation.

A useful tool in this context, given GP's ability to automatically select the fittest structures in the hypothesis space, is to inspect the occurrence frequencies of the function and operator symbols in the population. The evolution of symbol frequencies for two of the three primitive sets used in our example is illustrated in Figure 4.6. The frequency curves can be used as a guide to determine which symbols work best for a given problem and algorithm configuration. For example, the tendency for a symbol to become extinct after a number of generations could indicate that this symbol is unnecessary or even detrimental for a particular problem.

Nevertheless, this approach should be used carefully since it is also possible for symbols to become extinct for other reasons such as inappropriate algorithm configuration (population size, maximum number of generations, selection pressure, etc.). In some cases, a symbol may become useful only under specific conditions and although it might play an important role in achieving a good solution, if the conditions are not reached then its frequency curve might offer misleading information about its importance.

The reader may ask herself at this point why we were not able to rediscover the generating function in the second experiment. After all the hypothesis space in the second case contains the generating function as well as many equivalent expressions and therefore GP should be able to rediscover the optimal solution with an NMSE of zero. While we cannot give a definite answer to the question, there are several possible explanations. It could be that the size limit is too restrictive to allow GP to assemble the optimal solution, or that the population size is too small. Another potential cause could be that GP had difficulties recognizing the optimal solution. Note that the problem requires to find two sums for the two denominators and that the target function is not defined for x or y equal zero. This can be difficult for GP because even if it visits a solution candidate structurally similar to the generating function there is a high probability that evaluation leads to a division by zero which implies a low fitness value. Correspondingly, GP would not recognize that it has found a good model structure. This highlights again the necessity to use prior knowledge. If we know that the function we are looking for is only partially defined then we should make sure to handle this so that GP is able to evolve good solutions nevertheless. We already discussed a similar case in Chapter 3 and will provide more details on handling such cases in Chapter 6.

4.8   GP Dynamics

The characteristics of the hypothesis space search of EC methods are fundamentally different from local search (e.g., gradient descent). Local search methods search for better solutions by moving from the current point towards a local optimum e.g., following the direction of steepest descent. With EC methods, however, random properties are initially distributed over many solution candidates. The pieces of the puzzle, which when correctly assembled result in a high quality solution, are on the table from the beginning and it is the task of the evolutionary search process to bring together the properties distributed over the solution candidates. For SR, relevant parts of expression trees consisting of operators and operands – so-called *building blocks* – have to be combined in such a way that they result in an expression with low error for the training set. The evolutionary search allows to gain information about the importance of variables and mathematical operators by observing the dynamics of the search. The initially random distribution of operators and operands changes over time as the exerted selection pressure leads to proliferation of building blocks that are helpful and extinction of parts that are irrelevant

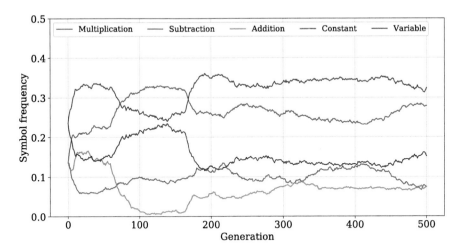

(a) Symbol frequencies recorded in the best run with primitive set
$(+, -, \times, \text{constant}, \text{variable})$

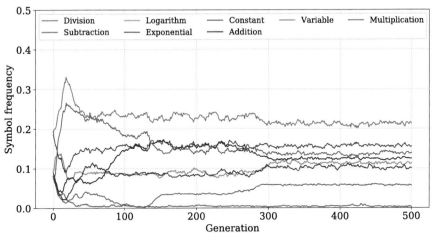

(b) Symbol frequencies recorded in the best run with primitive set
$(+, -, \times, \div, \exp, \log, \text{constant}, \text{variable})$

Figure 4.6: Evolution of symbol frequencies per generation for two primitive sets.

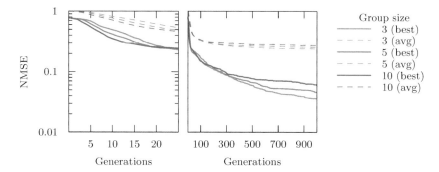

Figure 4.7: The selection pressure is controlled via the tournament selection group size and determines how quickly fitness improves. A high selection pressure can lead to more rapid improvement in the beginning, but in the long term a lower selection pressure can be more beneficial. The left panel is a zoomed view that shows the first 25 generations of the right panel, which shows average and best fitness values over 1000 generations.

to solve the problem. Plotting the empirical distribution of operators and operands or even larger building blocks in the population allows us to detect when this happens too quickly and too greedily (e.g., due to too high selection pressure) which leads to premature convergence.

4.8.1   Fitness

The observation of fitness over time already allows rough insights and conclusions about the convergence behaviour. If algorithm parameters are set too greedy (small population size, high selection pressure), this typically expresses itself in initially quick progress followed by early stagnation. If, on the other hand, selection pressure is not high enough, fitness improves only slowly but can reach a better value in the long term as shown in Figure 4.7. A plot of the best, average, and worst fitness in the population over time can be insightful.

Admittedly, such conclusions are rather superficial, since they only consider the fitness of solution candidates and the true causes for sub-optimal search dynamics of GP often lie on the genotypic level. We will discuss tracking diversity dynamics on the genotypic level in Section 4.8.4.1. However, first we discuss phenotypic properties of expressions that we may track over time to gain relevant insights.

4.8.2   Variable Relevance

GP has the potential to implicitly select relevant input variables. Even when we provide a large set with hundreds of potentially relevant variables the algorithm will evolve a model which contains only a small subset. This embedded feature selection step of SR has been proven to be highly competitive compared to other methods (Smits et al., 2005; Stijven et al., 2011). Additionally, we do not need to provide calculated features such as nonlinear transformations of variables (e.g., $\frac{1}{x}$), ratios ($\frac{x}{y}$) or interactions (xy) because GP can evolve these automatically. However, there are several facts that we need to consider when selecting input variables.

Perfectly correlated variables should be removed to reduce the size of the search space. It is fine to keep only one variable from the set of pairwise perfectly correlated variables. Be careful when removing highly correlated variables because even very highly correlated variables might provide relevant information when combined (e.g., x1 - x2, when x1 and x2 are highly correlated). For highly correlated variables GP may include one or the other randomly. This can lead to inconclusive results. Therefore, it can be beneficial to keep only one representative from a set of highly correlated variables.

Even though GP automatically selects the most relevant variables, we should still remove variables from the terminal set if we know they are irrelevant, again reducing the hypothesis space. An overly large input set increases the chance that irrelevant variables (spurious correlations) are included in the final model. GP may also include irrelevant variables in bloated models where they have only minimal effect on the output.

Population size should be adjusted based on the number of input variables to allow all variables, and many different nonlinear transformations and combinations of variables to occur within the genome.

It is not strictly necessary to scale variable values because GP should be able to identify scaling factors automatically. However, scaling may help the algorithm to identify good solutions. There is no conclusive answer to the question whether scaling of inputs is beneficial for GP-based SR. The effect is likely to be problem-dependent. It is worthwhile to make some initial experiments with and without scaling of variables (e.g., to the range 0..1) to see if there is a difference in solution quality.

It is possible to measure the importance of the variables by counting their occurrence in the expressions of the final population. For example we can analyze all solution candidates in the population and count how often each variable is referenced (Winkler, 2009). Additionally, we may weigh solution candidates by their fitness to put more weight on variables used in solution candidates with higher fitness (Kordon et al., 2006). If we determine variable relevance in this way, we are able to observe how variable relevance changes in the evolutionary process over the generations. At the beginning, we have no information about the relevance as all variables occur with similar frequencies. Later GP amplifies the frequencies of variables which occur in fitter individuals and the frequencies of less relevant variables will be reduced accordingly. Over time, variables which are not helpful will be removed completely from the population. If the process is allowed to run long enough we may also see random fluctuations in variable frequencies even though fitness stays more or less the same. This means that GP finds solution candidates of similar fitness

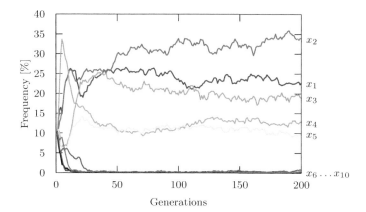

Figure 4.8: Relative variable frequencies for an exemplary GP run for the Friedman problem Equation (5.1).

using different sets of variables and is an indicator that some variables may be redundant or correlated.

Figure 4.8 shows the relative number of variable references (relative variable frequency) in the population over 50 generations for x_1 to x_{10}. The data is from one exemplary GP run on a simple function of ten variables (Equation (5.1) taken from Friedman (1991)). In this problem, the variables x_1 to x_5 are necessary to predict the dependent variable, whereas x_6 to x_{10} are not necessary. Figure 4.8 clearly shows that GP is able to quickly identify the relevant variables for this dataset. In Section 5.1.2.1 we discuss variable frequencies in more detail and describe other methods to determine variable relevance.

4.8.3   Model Complexity

The complexity of solution candidates is another property that is worthwhile to track and visualize. We identity two kinds of complexity:

- *Structural complexity* is the complexity of the expression or its tree representation. During the course of evolution the tree expressions may become very large (up to the established size limit) or may include redundancy in the form of bloat and introns.

- *Semantic complexity* is the complexity of the output of the expression (of the function). Increasing semantic complexity increases the danger of overfitting.

Structural complexity measures are relatively easy to calculate from the encoded expression. Examples are tree size, tree depth for tree-based GP, the number of variable references, the number of nonlinear functions, or the maximum number of variables in interaction. There are also structural complexity measures that estimate readability or explainability of expressions for example the recursive complexity measure (Kommenda et al., 2015) penalizes nesting of nonlinear functions, visitation length (Keijzer and Foster, 2007) includes tree shape and favors more regular trees.

Semantic complexity is harder to estimate as it requires to calculate the complexity of the output of an expression on finite training data or maybe even for the whole (infinite) input domain. Intuitively, a function is complex if it has many local optima or many points of high curvature. These properties are hard to calculate especially for multi-variate functions. Examples for semantic complexity are the nonlinearity of functions (Vladislavleva et al., 2009) or the number of turning points (or local optima) for uni-variate functions (Vanneschi et al., 2010).

Since size and complexity are properties that generally offer an evolutionary advantage via mechanisms such as genetic buffering, redundancy and neutrality, any available margin for growth – and complexity – will be taken advantage of by the process until the growth limits are reached. Nevertheless, we may spot important problems by analyzing size or complexity dynamics. For instance we often observe an extreme sudden drop in program sizes in the first few generations which implies a dangerous loss of diversity. This can occur if only small random solutions have high fitness in the initial population.

4.8.4 Diversity

From the point of view of selection, an individual's survival depends on whether or not its genotype is able to express a fit phenotype. Therefore, the mapping between genotype, phenotype and semantics are of interest for understanding the dynamics and performance factors of GP.

An important aspect in EC is the *locality* of the genotype-phenotype map (Rothlauf, 2006, 2016), which describes the map's ability to preserve neighborhood relationships across the genotype and phenotype spaces. When neighborhoods are preserved, the map is said to have high locality (e.g., a small change in the genotype will result in a small change in fitness). In general, high locality is a prerequisite for a successful search. However, optimization algorithms solving difficult or deceitful problems may exhibit better performance under low locality (Galvan-Lopez et al., 2012). Investigating locality requires appropriate distance measures for genotypes, phenotypes, and semantics, motivating the need for GP diversity analysis. In tree-based GP we directly operate on trees and our main concern is the locality of the mapping between trees and their outputs (semantics).

Population diversity is also tied to algorithm performance through the concept of *evolvability*, defined as a genotype's *potential* to produce adaptive change through recombination (see Section 4.1). Since this potential is not reflected in individual's fitness, potential issues may arise when genotypes with high evolvability are culled from the population before their genes get the chance to be combined with other genes into a better configuration.

In this context, measuring diversity and correlating its evolution with algorithm performance may lead to a better understanding of the dynamics of the evolutionary process. A plethora of diversity measures have already been developed for GP (Burks and Punch, 2018).

In the following we introduce practical genotype and semantic diversity measures that can be applied to GP. These measures are designed to be easy to use and only depend on standard algorithmic components such as a tree distance measure or a correlation measure (Winkler et al., 2016; Affenzeller et al., 2017; Burlacu et al., 2019).

4.8.4.1   Genotypic Diversity

Metrics for genotypic diversity are typically modeled after distance metrics such as the tree edit distance, which involve complex logic for dealing with associativity and commutativity and deciding subtree isomorphism. Furthermore, to calculate an average diversity score for a population of n individuals, one would need to perform $\frac{n(n-1)}{2}$ pairwise distance calculations (assuming a symmetric distance measure), incurring a large runtime cost. Due to their high computational requirements distance-based metrics have not been very popular with GP.

Algorithm 2 TREE HASH ALGORITHM

> **input:** An expression tree T
> **output:** The corresponding sequence of hash values

1: hashes ← empty list of hash values
2: **for all** node n in Postorder(T) **do**
3:      $H(n) ←$ an initial hash value
4:      **if** n is an internal node **then**
5:          **if** n is commutative **then**
6:              Sort the child nodes of n
7:          child hashes ← hash values of n's children
8:          $H(n) ←$ Hash(child hashes, $H(n)$)
9:      hashes.append($H(n)$)
10: **return** hashes

A practical method for avoiding the computational cost of complex tree distances is to employ a hashing approach for finding isomorphic subtrees (Burlacu et al., 2019). Using a tree hashing algorithm, the computation of the population distance matrix can be performed in two stages:

1. **Hashing**. Iterating over the population only once, each tree is hashed into a sequence of hash values corresponding to each node.

2. **Distance calculation**. Using the hash sequences obtained for each tree

in the previous step, tree distance is efficiently computing using array intersection.

The hashing approach replaces $\frac{n(n-1)}{2}$ calls to an expensive distance function between two trees with n calls to a tree hashing function and $\frac{n(n-1)}{2}$ calls to a comparatively much cheaper distance function between two node hash sequences.

A distance metric can then be defined using array intersection, for instance as the complement of the Jaccard coefficient, to obtain a measure in $[0, 1]$:

$$\text{GenotypicDistance}(T_1, T_2) = \frac{|H_1 \cup H_2| - |H_1 \cap H_2|}{|H_1 \cap H_2|}$$

where T_1, T_2 are two tree individuals and H_1, H_2 are their corresponding sequences of node hash values.

By replacing a complex tree distance with a much simpler array intersection, order-of-magnitude speed-ups can be achieved, making it feasible to use this type of genotypic distance for measuring diversity during GP runs. The procedures are shown as pseudo-code in Algorithms 2 and 3.

Algorithm 3 HASH-BASED TREE DISTANCE

---

input: Two sorted hash arrays H_1 and H_2
output: The genotype distance between trees T_1 and T_2

$i \leftarrow 0$, $j \leftarrow 0$, count $\leftarrow 0$
while $i < |H_1|$ **and** $j < |H_2|$ **do**
   if $H_1[i] = H_2[j]$ **then**
      count $\leftarrow$ count $+ 1$
      $i \leftarrow i + 1$
      $j \leftarrow j + 1$
   else if $H_1[i] < H_2[j]$ **then**
      $i \leftarrow i + 1$
   else
      $j \leftarrow j + 1$
return $1 - \dfrac{\text{count}}{|H_1| + |H_2| - \text{count}}$

4.8.4.2 Semantic Diversity

The argument for employing semantic diversity is that since selection acts on semantics or behaviour of individuals, it could be more worthwhile to promote behavioural diversity instead of structural diversity (Burke et al., 2004).

Metrics for semantic diversity take into account model output over the training data (so-called *sampling semantics*) and typically use a correlation measure to aggregate a diversity value. Correlation naturally corresponds to a similarity measure, from which a distance can be easily derived.

A semantic diversity measure between two individuals can be defined using the squared Pearson correlation coefficient between their corresponding responses:

$$\rho_{X,Y} = \frac{\text{cov}(X,Y)}{\sigma_X \sigma_Y}$$

In the situation where one of the models has a constant output, the denominator will evaluate to zero and, thus, we need to handle these special cases:

$$\text{SemanticDist}(T_1, T_2) = \begin{cases} 0 & \text{if } \sigma_{r_1} = 0, \sigma_{r_2} = 0 \\ 1 & \text{if } \sigma_{r_1} = 0, \sigma_{r_2} \neq 0 \text{ or } \sigma_{r_1} \neq 0, \sigma_{r_2} = 0 \\ 1 - \rho^2_{r_1, r_2} & \text{if } \sigma_{r_1} \neq 0, \sigma_{r_2} \neq 0 \end{cases}$$

Here, individuals T_1, T_2 have responses r_1 and r_2. According to the definition, the semantic distance is always in the interval $[0, 1]$. The distance is considered to be 1 when comparing a constant response with a non-constant response, and 0 when both responses are constant.

A potential weakness of correlation-based measures is that they follow the expected trend of models becoming "more similar" with the modelling target and consequently also with each other.

To illustrate how selection pressure affects population diversity (at both genotype and semantic level) we compare GP with tournament selection and GP with *gender-specific parent selection*[1] and strict *offspring selection* (OSGP) described in Section 4.9.1. Offspring selection is meant to reject deleterious genetic recombination events by accepting generated offspring into the new population only if they are (to a certain degree) at least as fit as one or both of their parents.

In terms of diversity, this two-stage selection mechanism causes a rapid proliferation of fit building blocks within the population, to the detriment of other less fit genes that are driven to extinction. Thus, the more effective and intense exploitation of the genes in the population is also accompanied by an accelerated rate of diversity loss. This effect can be observed by comparing Figures 4.9 and 4.10, where the evolution of diversity and fitness is shown for the two GP configurations.

To simplify the visualization, we illustrate diversity using the complemen-

---

[1]Typically, offspring selection GP uses a combination of proportional and random selection to select parents, meaning that one parent (the "female" parent) is selected proportionally based on fitness while the other parent (the "male" parent) is selected at random.

tary measure of *similarity*, defined as $1 - distance$:

$$\overline{\text{Sim}}_g = \frac{2}{N(N-1)} \sum_{i}^{N-1} \sum_{j>i}^{N} 1 - \text{GenotypicDistance}(T_i, T_j) \qquad (4.9)$$

$$\overline{\text{Sim}}_s = \frac{2}{N(N-1)} \sum_{i}^{N-1} \sum_{j>i}^{N} 1 - \text{SemanticDistance}(T_i, T_j) \qquad (4.10)$$

Here, $\overline{\text{Sim}}_g$ and $\overline{\text{Sim}}_s$ represent *mean similarities* calculated with respect to the genotypic and semantic distances. Since OSGP performs more evaluations per generation than GP due to the offspring selection step, the evolution of similarity and fitness over time is plotted with respect to the number of evaluated solutions so far. Without offspring selection, the same number of fitness evaluations is performed each generation.

Figure 4.9a shows that with tournament selection after an initial decrease, population diversity is maintained at a constant value, representing an equilibrium between diversity introduced by the recombination operators (mutation in particular) and diversity lost due to the action of selection. By contrast, with offspring selection population diversity keeps decreasing (as shown by the ascending curves in Figure 4.10a) and semantic diversity in particular goes towards zero. This is due to the correlation-based nature of the semantic diversity metric and the offspring selection step accepting child individuals that are semantically more similar to their parents.

In terms of average and best fitness shown in Figures 4.9b and 4.10b, the different dynamics between GP with tournament selection and offspring selection are seen as the relative difference between best and average fitness curves, f_{max} and $\bar{f}_i$. With tournament selection, since recombination carries an expected decrease in fitness, the increase of mean fitness per generation is strictly due to the selection mechanism. For these reasons a large difference between the mean population fitness and the fitness of the best individual is observed. With offspring selection, the requirement that recombination increases fitness (since fitness-decreasing recombination events are rejected in the offspring selection phase) leads to a steeper increase of the mean population fitness over time, up to a point where its value becomes very close to the fitness of the best individual. This intuitively points to the increasing difficulty of obtaining child individuals better than their parents as the search progresses, making it suitable as a dynamic termination criteria for the algorithm (when it becomes no longer possible to obtain better children from the existing parents).

More detail about population diversity can be gathered from density plots (Figures 4.11 and 4.12) that show the distribution of *point similarity* values in the population. In this case, point similarity is given by an individual's mean similarity with the rest of the population (in other words, point similarity values are the row means of the population similarity matrix). Using this approach, we observe different distributions underlying mean values for genotype and semantic similarity. For example, Figure 4.11 shows that we have greater

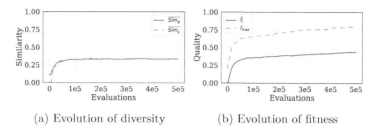

(a) Evolution of diversity   (b) Evolution of fitness

Figure 4.9: Evolution of population diversity (left) and fitness (right) with tournament selection. The similarity curve quickly plateaus and there is a large gap between best and average fitness.

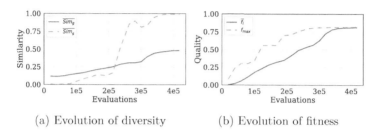

(a) Evolution of diversity   (b) Evolution of fitness

Figure 4.10: Evolution of population diversity (left) and fitness (right) with offspring selection. Offspring selection decreases diversity very strongly and the best and average fitness converge.

variation in terms of the semantics with tournament selection, whilst Figure 4.12 shows that semantics quickly become similar as a result of the offspring selection step. In both cases genotype similarity remains around the same mean value of ≈ 0.3.

4.9   Algorithmic Extensions

In this section we discuss several extensions of GP which are motivated by specific needs that arise in applications.

4.9.1   Brood Selection and Offspring Selection

This class of selection methods are designed with the goal of increasing the fitness of the offspring generation, making better use of the existing genetic material in the population and reducing the occurrence of deleterious mutations.

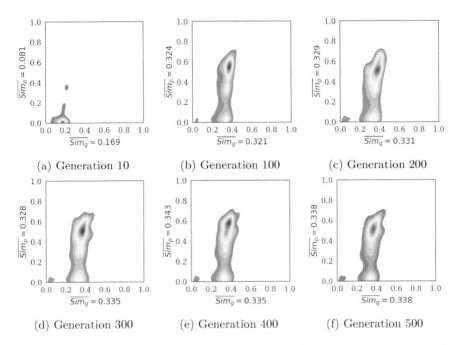

Figure 4.11: Distribution of pairwise similarities (genotypic vs. phenotypic) between individuals in a GP process (with tournament selection) solving the Poly-10 problem.

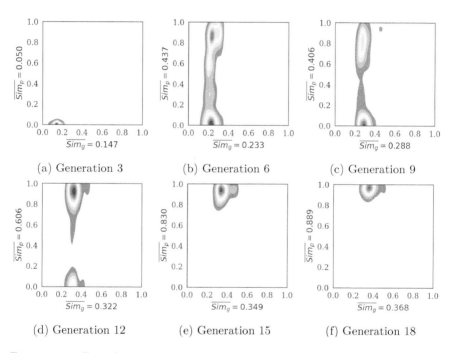

Figure 4.12: Distribution of pairwise similarities (genotypic vs. phenotypic) between individuals in a GP process (with offspring selection) solving the Poly-10 problem.

Their main operating principle is to retry the recombination process for a predefined number of times or until a satisfactory child individual is obtained.

Brood selection (Altenberg, 1994a) generates a "brood" of offspring from two parents and then keeps only the most fit individual. The goal is to shield the evolutionary process from deleterious changes and promote exploratory behaviour. The balance between exploration and exploitation can be controlled by means of a user-defined brood size parameter. A larger brood size will offer a better change of returning a fit offspring but will be more computationally expensive. Due to the fixed brood size, the same amount of child individuals (equal to the brood size minus one) are thrown away for every successful selection step. Since all of the generated offspring must be evaluated regardless of whether they are the best in the brood, brood selection GP is typically slower than a regular tournament selection GP by a fixed factor (equal to the brood size).

However, despite the increased computational cost, brood selection often offers superior performance due to the better exploratory behaviour of the search. For example, Tackett and Carmi (1994) show that increasing the brood size gives better performance than an equivalent increase in population size in standard GP.

Offspring Selection (OS) (Affenzeller et al., 2009) attempts to make better use of existing diversity by rejecting deleterious recombination operations. The generated offspring are accepted into the population only if their fitness satisfies a performance criterion.

Similar to upward-mobility selection (Altenberg, 1994b), OS accepts or rejects new offspring based on an interpretation of what "better" or "fitter" might mean at a given moment in time: better than one parent, both parents, or some in-between value. Successful recombination events are those which succeed at assembling together genes from parents to produce an offspring that fulfills the performance criterion. It also borrows ideas from simulated annealing (Kirkpatrick et al., 1983), introducing a *comparison factor* parameter for controlling the acceptance criterion:

$$\text{Success} = \begin{cases} \text{true} & f_{\text{offspring}} > \min(f_{\text{parent}_0}, f_{\text{parent}_1}) + c \cdot |f_{\text{parent}_0} - f_{\text{parent}_1}| \\ \text{false} & \text{otherwise} \end{cases}$$

$$(4.11)$$

According to Equation (4.11) offspring produced via recombination are accepted if their fitness surpasses a threshold between the fitness of the weaker and better parent.

By using a continuously increasing comparison factor c (similar to the temperature in simulated annealing), OS enables the evolutionary search to perform a wider exploration of the solution space in the beginning and a greedier search towards the end of the run.

The general workflow of GP with OS is shown in Figure 4.13. A simplified algorithm model may be obtained when both the *success ratio* and the *comparison factor* parameters are set to one. A success ratio of one means that

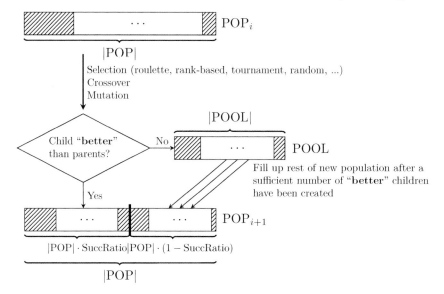

Figure 4.13: The OSGP extension.

100% of the new generation must be filled with individuals from successful recombination events, while a comparison factor of one means that the fitness threshold for success is the fitness of the better parent. This simplified OS procedure denoted *strict* offspring selection is shown in Figure 4.14.

OS actively promotes evolvability by selecting for chromosomes which can be recombined successfully. However, this comes at the cost of a considerably increased search effort due to the additional fitness evaluations of offspring that get rejected, raising the question of whether or not the extra computational resources would not be better spent on larger population sizes, as observed by Altenberg (1994b) in his discussion of upward-mobility selection.

Regardless of whether it provides better performance, OS can be convenient because it provides with an estimator for evolvability and (offspring) selection pressure. We can calculate offspring selection pressure for each generation by calculating the ratio of successful and unsuccessful recombination events over population size. If we need to generate and evaluate O offspring to fill the population of size N the offspring selection pressure is $\frac{O}{N}$. This measure is anti-correlated to evolvability. We have high evolvability when it is easy to generate successful offspring at the beginning and low evolvability when offspring selection pressure is high. OS uses the available diversity and squeezes out diversity over time until almost all diversity and evolvability is lost. Accordingly, the offspring selection pressure will increase as diversity is lost which gives an indication of algorithm convergence. Especially, a sudden increase in offspring selection pressure indicates that the algorithm has converged and can be

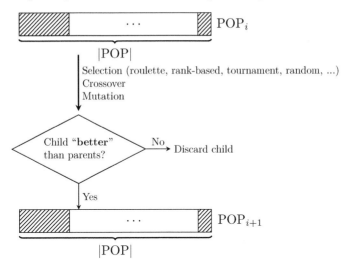

Figure 4.14: The *strict* OSGP extension.

stopped as it becomes very difficult to find even better solution candidates. We found a maximum selection pressure of 50 or 100 to work well.

OS can also free us from choosing a parent selection method because it even works well with random parent selection. However, the algorithm is less greedy in this case.

4.9.2 Age-layered Population Structures

In age-layered population structures (ALPS) (Hornby, 2006) diversity management is improved by restricting competition and breeding to individuals of similar age. Individuals are split into different age groups or *layers* whereby the bottom layer consisting of the youngest individuals is regularly replaced with randomly generated ones. In this way, the algorithm is never completely converged and the bottom layer acts as a steady diversity generator. By limiting competition to individuals of similar age (and therefore fitness) the diversity is kept high. Overall, ALPS introduces the following new evolutionary rules:

- Competition takes place exclusively between individuals in the same age layer or in the layer immediately below them.

- Each age-layer has an age limit for the individuals such that aging individuals that exceed the limit are moved into the next higher layer.

- Individuals that are created through recombination take the age value of the oldest parent plus one.

- Each generation in which an individual is used as a parent (at least once) to create an offspring, its age is increased by one.

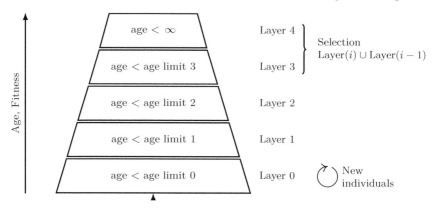

Figure 4.15: Age-layered population structure (ALPS).

- The bottom layer is regularly replenished with randomly generated individuals.

- The top (last) layer can have individuals of any age.

The age-layered population structure can be beneficial for dynamic problems such as SR problems where new data is continuously arriving in a data stream. When the environment changes and the previously fittest individuals lose their ability to survive, the algorithm can propagate new individuals which are better adapted to the new environment from the bottom layer towards the top. This makes ALPS particularly suitable for problems that dynamically transition between states or operating regimes with different optima.

4.9.3   Multi-objective GP

There are several generic options of combining multiple fitness values. These options are discussed frequently in the context of multi-objective optimization which is mainly concerned about combining multiple potentially competing objectives. The options are:

- Weighted sum of objectives: is often the easiest way to combine multiple objectives. Weighting is necessary either to bring the objectives to a common scale or to put more emphasis on objectives with a higher weight.

- Product of objectives: similar to a weighted sum but used when the overall fitness is a product of individual factors. An individual has a high fitness only if all fitness factors are high. Calculating an index of competing objectives (e.g., the fraction of error over length) is another example of this approach.

- Constraints (hard or soft): with constraints we set an acceptance threshold for some fitness dimensions and optimize only the other fitness dimensions.

This is useful in cases where the order of individuals is irrelevant once they reach the threshold value. To apply a hard constraint in GP-based SR we can simply set the fitness to the worst possible value. Soft-constraints are implemented by adding a penalty term to the fitness value. The penalty term can be calculated for instance via a nonlinear hockey-stick or hinge-function that is zero up to the threshold and then linearly increasing. A weight can be used to balance the fitness contribution of the main objective and the constraint violation.

- Multi-objective algorithms: these algorithms do not combine the individual fitness dimensions and instead try to find a set of Pareto-optimal solutions. Solutions within the Pareto-set are considered as equally useful. Multi-objective algorithms should be used if we explicitly aim to find many solutions that are spread well within the sub-space of Pareto-optimality.

- Hybrid: for instance we can use bi-objective optimization whereby using the error as one objective and a weighted sum of all other fitness dimensions as the second objective.

Optimization with many objectives is difficult and an area of active research, but for two or three objectives, several evolutionary algorithms have been proposed that often work quite well. We recommend to try to reduce the number of objectives as far as possible e.g., by using weighted combinations of multiple objectives, hard or soft-constraints. Multi-objective optimization is helpful mainly when the goal is to find many solutions with different tradeoff between the objectives.

As previously discussed, there is a need to limit the size of GP individuals because of GP's tendency to bloat which would lead to overly large, hard to understand, and potentially overfit solutions.

Multi-objective GP has the capabilities to promote parsimony or interpretability, and goodness-of-fit of regression models directly as orthogonal fitness objectives. In this scenario the goal is to find a set of solutions from which we may select a final solution which represents the best compromise. The concept of Pareto-optimality is at the core of multi-objective optimization. The set of Pareto-optimal solutions contains all solutions that are better than all other solutions for at least one objective. For example if we minimize size and error of regression models, the Pareto-optimal solution set will contain small models with higher error as well as larger models with lower error. These two objectives are typically diametric to each other which requires us to optimize for both objectives.

In general, multi-objective evolutionary algorithms (MOEAs) deal with competing objectives which have to be optimized simultaneously. The optimization goal is to obtain a Pareto front of evenly distributed solutions across the objectives from which the most suitable compromise might be selected. Several approaches for obtaining the Pareto front are possible.

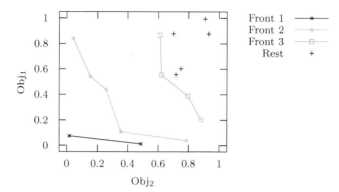

Figure 4.16: In multi-objective optimization the population can be ordered into multiple subsets (fronts). The plot shows an example for two objectives that should be minimized. The best solutions which are not dominated by any other solution are in front 1. All solutions within a front are equally ranked.

4.9.3.1 Approaches Based on Pareto Dominance

A majority of MOEAs employ the formalism of Pareto dominance to impose a partial ordering on the individuals (as shown in Figure 4.16) in the recombination pool and to decide which individuals participate in recombination.

In the formalism of Pareto dominance, two fitness vectors $y, z \in \mathbb{R}^M$ can have one of the following dominance relationships:

$$\text{equality} \qquad y = z \Leftrightarrow y_i = z_i, \forall i = 1, ..., M \qquad (4.12)$$

$$\text{strong dominance} \qquad y \prec z \Leftrightarrow y_i < z_i, \forall i = 1, ..., M \qquad (4.13)$$

$$\text{weak dominance} \quad y \preceq z \Leftrightarrow y_i \leq z_i, \forall i = 1, ..., M \text{ and } y \neq z \qquad (4.14)$$

Thus, if $y \preceq z$ then we say that y *dominates* z (z *is dominated by* y). If $y \npreceq z$ and $z \npreceq y$ then y and z are *mutually non-dominated*. A *Pareto front* represents a set of mutually non-dominated solutions.

The procedure of dividing a population of solution candidates into corresponding Pareto fronts based on dominance is known as non-dominated sorting. An overview of the most well-known MOEAs based on non-dominated sorting is given below.

The *Non-dominated Sorting Genetic Algorithm* (NSGA) performs pairwise Pareto dominance comparisons to divide the population into Pareto fronts. In its first version, NSGA (N. and Deb, 1994) did not employ any elitism and used a rather inefficient non-dominated sorting algorithm with $O(MN^3)$ asymptotic complexity, where N is the population size and M is the number of objectives.

The next iteration called NSGA-II (Deb et al., 2002) improved the complexity of non-dominated sorting to $O(MN^2)$ and introduced elitism via plus selection (the best individuals from the union of parents and offspring are selected for the next generation). Additionally, it introduced the notion of crowding distance within the same Pareto front as a tie-break mechanism to help select more diverse individuals.

Finally, a subsequent adaptation called NSGA-III (Deb and Jain, 2014; Jain and Deb, 2014) aimed to overcome the inherent difficulties of dominance-based MOEAs in handling many objectives, stemming from the fact that with an increase in the number of objectives, a large fraction of the population becomes non-dominated. NSGA-III promotes population diversity by adaptively maintaining a set of well-spread reference points that properly-normalized population members are associated with in each generation.

The *Strength-Pareto Evolutionary Algorithm* (SPEA) uses strength defined as an individual's dominance ratio over the rest of the population to compute fitness. The first iteration of SPEA (Zitzler and Thiele, 1999) defined fitness as the sum of the strengths of all individuals dominated by the current solution. Elitism is implemented with the help of an external archive storing all non-dominated solutions found so far. The next generation is selected from the union of the current population and the external archive. When the external archive gets full, pruning is performed with the help of an agglomerative clustering method.

Subsequently, SPEA-2 (Zitzler et al., 2001) introduced a more fine-grained fitness assignment strategy where for each individual both dominating and dominated solutions were taken into account. Agglomerative clustering was replaced with a nearest-neighbor search as a truncation technique, in order to prevent boundary solutions from being removed during the pruning of the external archive.

Hypervolume-based algorithms assign fitness to individuals based on the hypervolume indicator. The hypervolume indicator is the only indicator known to be stricty monotonic with respect to Pareto dominance, guaranteeing that the Pareto-optimal front achieves the maximum hypervolume possible. This property is particularly useful in many-objective scenarios, however, the problem of computing the hypervolume is NP-complete and no polynomial-time algorithhm is available. For this reason algorithms such as HypE (Bader and Zitzler, 2011) estimate the hypervolume using Monte-Carlo simulation to avoid high runtime costs.

4.9.4   Alternative Encodings: Linear and Graph GP

Linear GP encodes solution candidates as linear sequences of instructions, moving the representation closer to the machine level, and using linear data structures and registers for evaluation. The simplicity of the encoding gives up some flexibility in return for better runtime performance, usually orders of magnitude faster if the representation can be executed immediately by the machine without interpretation overhead. Notable examples of GP systems using a linear encoding are Linear GP (Holmes and Barclay, 1996), Stack-based GP (Spector and Stoffel, 1996), and Machine-code GP (Nordin et al., 1999). Push is a language for linear stack-based GP with several up-to-date implementations of GP systems based on Push (Spector et al., 2005). Push is designed for general program synthesis and provides many primitives including iteration, recursion, and even functions and modules (Saini et al., 2022).

Graph-based GP represents individuals as directed a-cyclic graphs, allowing the algorithm to evolve structures such as neural networks or finite-state automata, or to achieve more compact representations by exploiting modularity within the structure of the computer programs (Schmidt and Lipson, 2007). The main difference to the tree representation is that in a graph representation a calculation result from a sub-graph can be used in multiple up-stream paths.

These advantages come, however, at the cost of a more complicated implementation of the GP operators such as crossover and mutation.

Parallel Distributed Genetic Programming (PDGP) (Poli, 1996) and Cartesian GP (CGP) (Miller, 1999) are two well-known examples of graph-based GP. CGP is both grammar-guided and graph-based as it uses a linear representation which is expanded into a graph before evaluation, using rules which map integer tuples from the linear representation to graph nodes placed on a 2-dimensional lattice. Special care must be taken at the level of linear GP operators that the created offspring can be expanded into valid graph representations. Eureqa, which used to be a well-known software for SR, used a graph-based encoding as well (Schmidt and Lipson, 2009).

4.9.5   Restricting Expressions: Syntax and Types

We mentioned the requirement for closure of the primitive set which essentially means that all elements of the primitive set can be combined by GP freely because the inputs and outputs of all operation are of the same type. In Koza-style GP, closure is necessary because there is no syntax that restricts how symbolic expression trees are assembled. When using GP to evolve more complex programs one quickly recognizes that it would be great to have a way to restrict the structures assembled by GP. For example let's assume we would like to evolve a program for sorting arrays. The program takes an unsorted array and sorts the values in-place. The primitive set for such programs must contain at least operators for array-access, for comparison of two array elements, integer operations to manipulate array indices, a way to execute multiple operations in sequence, and conditional as well as loop or recursion operators. This primitive set does not fulfill the closure property (e.g., integer operations require integer arguments but the array access operator returns the array element type).

Even for SR – a problem much less complex than evolving a sorting program – it can be useful to restrict solutions. For instance we may want to limit the complexity of solutions by disallowing nesting of nonlinear operations, or we may want to fix a part of the structure if we already know parts of the solution (e.g., from earlier GP runs). Grammar-guided and strongly-typed GP allow such restrictions.

A formal language is a potentially infinite set of symbol sequences which can be defined via a grammar with a finite set of production rules. Within this formalism grammars allow to recognize and validate whether a given symbol sequence is part of a language. The theory of formal languages is the basis for the design and implementation of programming languages and compilers.

Grammar-guided GP systems use a formal grammar to evolve only valid solution candidates. There are different ways how this can be accomplished. One way is to use a linear encoding which is then decoded using the grammar rules to produce the phenotypes. Another way is to use a tree encoding and grammar-aware operators for tree initialization and recombination to ensure that only valid trees are produced. Linear encodings use bit strings or integer vectors while tree encodings use derivation trees. The main variants of grammar-guided GP with linear encoding are Grammatical Evolution (GE) (Ryan et al., 1998), Cartesian GP (CGP) (Miller, 1999), and Gene Expression Programming (GEP) (Ferreira, 2002). Variants of grammar-guided with tree encoding are CFG-GP (Whigham, 1995), LOGENPRO (Wong and Leung, 1997), and GGGP (Geyer–Schulz, 1996).

Tree-adjoining-grammar-guided GP (TAG-GP) (Hoai et al., 2003) is a subsequent variant of grammar-guided GP where the context-free grammar is replaced by a tree-adjoining grammar, with the goal of facilitating the introduction and preservation of language bias in the evolutionary process.

TAG-GP provides a separation between derivation and derived trees and a more natural genotype-phentype map. Genotypes are represented by derivation trees in a lexicalized tree adjoining grammar (LTAG). An LTAG requires that each elementary tree has at least one terminal node. Each node contains a list of lexemes for substitution with the elementary tree at that node. The main operation is adjunction.

Since the set of languages generated by TAGs is a superset of the context-free languages, TAG-GP represents a more general form of grammar-guided GP. The authors demonstrate the successful application of this approach in the Boolean domain (6-multiplexer problem).

Phong et al. (2013) further introduce relational semantics and show that a crossover operator equipped with this knowledge can outperform standard GP on a collection of SR benchmarks.

Strongly-typed GP (STGP) (Montana, 1993) enforces data type constraints to function arguments by means of strong typing, using generic functions which act as templates for classes of strongly-typed functions.

Genetic operators such as crossover, mutation, or tree initialization are modified to take into account type constraints for each node. This increases the complexity of the operators and their runtime overhead. Montana (1993) reports successful application of STGP for some moderately complex problems involving multiple data types, but concedes that its runtime complexity makes it infeasible for larger problems.

In conclusion, both approaches are viable to restrict how GP is allowed to combine primitives to programs. However, this functionality also increases the complexity of GP implementations and their parameterization. When using these systems we must be aware that any additional restriction on the individuals will effect evolvability and diversity. For instance with grammatical evolution it can be difficult to design a grammar which allows enough diversity in the randomly created initial population. As discussed in Section 4.8 we recommend to monitor solution size and diversity especially when using grammar or type constraints.

4.9.6   Semantics-aware GP

Semantics-aware GP methods take the semantics of individuals or parts of the representation of individuals into account to improve GP performance. For example a semantically-aware crossover (Nguyen et al., 2011; Uy et al., 2011, 2013) considers semantics to increase the likelihood to successfully combine beneficial properties from both parents. Semantics-aware methods have been studied for a long time (Beadle and Johnson, 2008, 2009) and a good survey is provided by Vanneschi et al. (2014a).

Geometric semantic genetic programming (GSGP) (Moraglio et al., 2012) uses genetic operators that exploit the geometric properties of semantic space. It uses the notion of semantic distance (a metric in semantic space), where a program's semantics is defined as its output over a set of fitness cases. This makes the geometry of the semantic space a cone where the apex is the global optimum. The convexity of the semantics space allows to create individuals with improved fitness as a linear combination of two parents.

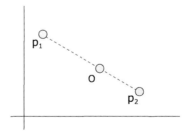

Figure 4.17: Offspring o produced by geometric semantic crossover of parents p_1 and p_2 (Vanneschi et al., 2014b).

A geometric semantic operator is considered *effective* if it avoids producing programs that are semantically equal to already considered programs. The simplest semantic geometric crossover produces a new offspring o from two selected parents p_1 and p_2 by the use of a randomly created program p_r, whose outputs must be in the range $[0, 1]$.

$$o = (p_r \times p_1) + ((1 - p_r) \times p_2)$$

This naive approach of creating children as linear combinations of their parents leads to an exponential growth of program size because the whole representation of both parents is copied to the children. While a smart implementation can circumvent reevaluation of the parts copied from parents (Vanneschi et al., 2013) to prevent an explosion of computational effort, the final solutions would still be huge. Therefore, in practice linear combination of parents is problematic, leading to the development of alternative operators that are approximately geometric (Krawiec and Pawlak, 2013b) leading to the concept of *semantic backpropagation* (Krawiec and Pawlak, 2013a; Wieloch and Krawiec, 2013).

Empirical evidence shows improved performance for approximately geometric crossover (Pawlak et al., 2015) for SR problems. These semantic operators are described as *competent* operators (competent initialization, selection, recombination, mutation). Competent semantic GP was shown to be successful in the Boolean and SR problem domains (Pawlak and Krawiec, 2018), indicating that GP operators designed with semantic and geometric aspects in mind are beneficial for the overall search performance.

The main idea of *behavioural program synthesis* (Krawiec, 2016) is to use data that is generated in the execution of GP individuals to improve GP performance. In many GP systems, a scalar fitness value is generated as an aggregate value over multiple test cases and selection uses only the scalar value. Through the aggregation a lot of information is lost. For instance if we look at the error for individual test cases, we may prefer to combine parents which are complementary to each other (low error on different test cases) instead of parents with low error on the same test cases. This is also the motivation for lexicase selection as discussed in Section 4.4. Going further the whole execution trace containing intermediate results produced when executing GP individuals can provide valuable information to guide the GP process. Krawiec proposes to combine multiple *search drivers* that can be calculated from the execution traces for behavioural GP.

4.10   Conclusions

In this chapter we introduced the general concepts for evolutionary computation methods and gave an introduction to GP. SR has been introduced in the foundational GP literature and even today, thirty years later, GP is still a popular and effective method for SR. In this time many algorithmic extensions to the classical tree-based GP framework have been developed which improve performance and thus allow to find solutions even for difficult SR problems. Especially, the recently actively researched area of semantics-aware operators for GP is notable. Semantics-aware GP can give us a clearer picture of semantic diversity in the population and uses information-rich semantics for selection and recombination instead of scalar fitness values. The outlook for further improvements to GP is positive as there are still many open topics to be solved. Additionally, the implicit parallelism of population-based methods is well-aligned with developments in computer hardware.

At first glance, EC methods seem to be nothing more than glorified random search. However, this is not entirely accurate: EC methods perform a *guided* random search. As we have shown in this chapter, the concepts evolvability, diversity, and selection pressure provide a intuitive understanding of the EC process and using these concepts we are able to tune parameter settings specifically for the problems we try to solve.

A key for understanding and correct parameterization of the process is to analyze the dynamics of population properties such as the distribution of fitness values and complexities, selection pressure, and diversity. Visualizations of these properties are especially useful. With this understanding of the internals of EC methods we are certain the reader will be able to use GP and interpret the results with confidence.

In the next chapter we give more details on the analysis and postprocessing of SR models produced by GP and describe methods for model interpretation, validation, and selection.

4.11   Further Reading

The series of books on GP by John Koza (Koza, 1992, 1994; Koza et al., 1999, 2003) can be recommended to get an impression of the history of GP developments. The books also contain many examples of successful applications of GP. *Genetic Algorithms in Search, Optimization and Machine Learning* by Goldberg (1989) is the primary reference for genetic algorithms. The *Field Guide to Genetic Programming* (Poli et al., 2008a) can be recommended for readers who just started to use GP. Its strength is the huge bibliography with relevant references. In this context, the GP Bibliography[2] curated and maintained by Bill Langdon is an invaluable resource when searching for specific algorithm extensions or applications of GP. Our own book *Genetic Algorithms and Genetic Programming* (Affenzeller et al., 2009) has more in-depth content on OS and related algorithm extensions as well as on real-world applications.

The most important conferences are the *Genetic and Evolutionary Computation Conference* (GECCO), the Evo* conference (especially the sub-conference EuroGP), the *Parallel Problem Solving from Nature Conference* (PPSN), and the IEEE Congress on Evolutionary Computation (CEC). All these conferences

[2]http://gpbib.cs.ucl.ac.uk/

have a rigorous review process (double-blind in the case of GECCO and Evo*) and accept only high-quality papers. The proceedings are available online and listed in the GP Bibliography.

The book series on *Genetic and Evolutionary Computation* provides regular updates about recent GP research and has a yearly edition with chapters from participants of the Workshop on *Genetic Programming in Theory and Practice* taking place in Ann Arbor or East Lansing, Michigan (for example Banzhaf et al. (2020), Banzhaf et al. (2022), Trujillo et al. (2023)).

5

Model Validation, Inspection, Simplification, and Selection

With its ability to generate interpretable and explainable nonlinear models, SR has the potential to significantly contribute to the growing field of explainable artificial intelligence and scientific machine learning.

In this chapter, we will discuss several methods which allow a more detailed analysis of individual models and their outputs. We will also discuss possibilities to inspect, explain, and simplify SR models. Our main focus is to make SR results understandable to domain experts.

The main advantage of using GP is the robustness given by its stochastic component, which allows many other heuristics and optimization paradigms (for example, hybridization with local search algorithms, inclusion of soft and hard constraints, integration of prior knowledge into the process) to be combined under the evolutionary paradigm such that modelling goals can be flexibly addressed. However, as a non-deterministic method, GP has the tendency to generate wildly different models even for similar training data. The reason is not so much the usage of a pseudo-random number generator, as we could easily fix the sequence of random numbers, but instead the multiplicity of potential solutions having effectively the same goodness-of-fit and the multiplicity of representations for the same solution. GP-based SR should be understood as a *hypotheses generating machine* instead of an algorithm that generates the "best model". In this respect the diversity of models with similar accuracy that are generated by SR is a strength. The downside is that we need to select one or maximally a handful of models from this set of candidates because we often want to use only a single or a few models in the application. Different models for the same problem can be analyzed to find commonalities between these models. For instance we may analyze which input variables have been selected by GP in all or most models or whether certain expressions reoccur in multiple models.

Model analysis serves first and foremost the purpose of validating the modelling process and answering the question of whether the obtained model indeed captures the essential properties of the studied system, and is therefore able to generalize. Secondly, model analysis serves the purpose of knowledge discovery, gaining new insight and understanding previously unknown aspects pertaining to the modeled system's structure or behaviour.

Model selection is the task of choosing a final model from a set of potential models. Typically, we have multiple candidate models with similar goodness-of-fit and we have to select a single one. Through careful analysis and with the help of statistical indicators we can reduce the probability of selecting a bad model.

Model analysis or validation and model selection go hand-in-hand and are mainly manual tasks that require expertise in statistics and the interpretation of models as well as expertise in the application domain. Model selection should be driven by the scientific plausibility and applicability of the model and should be supported by statistical analysis for the comparison of multiple models (Burnham and Anderson, 2003). For this task, we need analysis tools that enable us to thoroughly understand and compare different models. First and foremost, all candidate models should fit the data. To select from equally well-fitting models we have to understand the individual models in detail, to establish whether they are scientifically plausible. If multiple models are equally plausible, Occam's razor tells us to prefer the less complex model. Statistical measures such as the Akaike information criterion (AIC), or the Bayesian information criterion (BIC) allow us to quantify the likelihood of models and compare relative likelihoods between models. These measures can also be helpful for model selection, especially when we do not have enough data to use a part of it for model validation.

Prior to model analysis and selection we also have to consider model selection within the algorithmic part of the modelling process. For example GP explores millions or billions of candidate models and has to select candidate models within its main loop. Additionally, GP has a final step of automatic model selection because one or a few models have to be returned to the user. Finally, we usually want to select a single model from multiple candidates returned by a GP run or from multiple GP runs. This selection task is the main focus of the following sections.

We will cover techniques for model validation first and afterwards give hints how to select the final model from a set of validated candidates.

5.1   Model Validation

We already discussed error measures to quantify the goodness-of-fit in Chapter 2. These error measures are useful to compare models, but they only give aggregate information about the goodness-of-fit. Models are often presented without even rudimentary validation especially in the SR literature. Often, only aggregate errors on a testing set are given, without more detailed analysis of the model. This indicates that there is a lack of awareness of the importance of the topic and further stresses the need to cover this topic here. Often, finding a well-fitting predictive model comes secondary to other goals such as:

- The ability to understand and explain model behaviour

- Scientific plausibility of the model

- Whether the model is generalizable and capable of extrapolation

- Boundedness and safe operation under all circumstances

- Efficiency of calculating predictions or computational effort required for training the model

These objectives require a model representation flexible enough for learning algorithms and at the same time simple enough for the user to interpret and understand. A mathematical representation of models in the form of expression trees fits both these requirements, with the observation that interpretability will still be contingent on model size and complexity, and these two attributes must therefore be controlled during the learning process.

The SR modelling workflow involves a certain amount of experimentation with regard to specific GP parameterizations and extensions that may improve models. Considering the stochastic nature of the modelling process, a large number of models is typically generated for each configuration. It is therefore necessary to process this set of models and extract a subset that best fulfills practical criteria imposed by a given application domain.

In this section we discuss model selection and validation techniques that help identify the most appropriate models from a collection of models according to a set of criteria such as model performance, generalization ability, interpretability, or physical plausibility.

We distinguish here between *explainability*, representing the extent to which the internal workings of a model can be explained to domain experts, and *interpretability*, representing the extent to which we can establish physical meaning for components of the model such as individual parameters or selected terms. The following section focuses on explaining models.

5.1.1 Visual Tools

Different forms of visualization of the model output are immensely useful for model validation. Using visualization we can understand the predictions as well as the inner workings of the model. Visualization helps to analyze the output for systematic deviations or implausible predictions. Additionally, visualization tools can also help to compare the goodness-of-fit of multiple models in more detail.

The basic visualizations for model validation are scatter plots of target and predicted values, probability or quantile-quantile-plots, and residual plots (Draper and Smith, 1998). Additionally, partial dependence plots or intersection plots help to understand main effects captured by the model for validation.

The basic steps and visualizations used for model validation are common among regression methods including SR. We cover the techniques that we

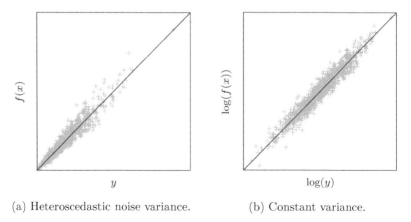

(a) Heteroscedastic noise variance.   (b) Constant variance.

Figure 5.1: Example scatter plots where the noise variance is correlated with the target variable y before and after transformation.

found most useful for SR. Further details can be found in one of the many good books on nonlinear regression modelling, see for example Bates and Watts (1988); Draper and Smith (1998).

One of the first plots to check for any regression model is a *scatter plot* of the predicted values over the target values. Alternatively, for time series models or when there is a natural order of observations in the dataset, a line chart of the target and predicted values can be useful.

A simple scatter plot as shown in Figure 5.1a can give us a lot of information about the validity of the model. For a model with perfect prediction (MSE= 0), all points would lie on the diagonal of the scatter plot. In the scatter plot we can easily detect systematic bias of the model, outliers with a strong impact on the model, and heterogeneous variance (heteroscedasticity) of the target variable.

The data shown in this plot is especially constructed to demonstrate heteroscedasticity. This means that there is a correlation of the noise variance with the target variable or with input variables. Heteroscedasticity is problematic when fitting with the assumption that errors are distributed independently and identically which means that all observations are weighted equally when fitting the model. If the scatter plot shows high heteroscedasticity then a model fit with least squares is affected more strongly by the observations with high noise variance. There are several ways in which this can be handled including: weighted fitting, nonlinear transformation of the target variable, using a different objective function (e.g., mean of relative errors), or using a different error model leading to a different likelihood function. If we know the standard errors of measurements we can simply account for the different measurement errors in the likelihood function. In this way, less precise observations have a smaller weight in the parameter estimation procedure than more precise measurements. Alternatively, nonlinear transformation of the target values may allow to homogenize the variance and to continue with a simple least squares approach. Target variables which can only be positive can be transformed using the square root or the logarithm. A more general transformation is the Box-Cox transformation $y' = (y + c)^\lambda$ with parameter $\lambda \in \mathbb{R} \setminus 0$ and $y' = \log(y + c)$ for $\lambda = 0$, which includes square root and logarithm as special cases. The parameter c in the Box-Cox transformation is necessary when we want to shift the target values to a certain range. The goal of the transformation is to stabilize residual variance over the full range of the target variable. Figure 5.1b shows the effect of the log-transformation for the constructed example shown above. This is a perfect example of identically distributed errors. Of course, for real datasets it is often much harder to judge whether the residual distribution is homoscedastic, especially with small datasets or low signal-to-noise ratios.

In the scatter plot we can see if the errors are identically distributed over y which is one of the assumptions usually made in regression. Another assumption in fitting concerns the distribution of errors. For example, in regression tasks we often assume that errors are normally distributed. This is also the assumption underlying least-squares fitting which is most common in SR literature, even though SR is not limited to this setting. The probability plot or often called *quantile-quantile plot (short QQ-plot)* as shown in Figure 5.2 can be used to check if the empirical distribution of the residuals of the model are close to the assumed error distribution. In least squares fitting the empirical distribution of the residuals of a well-fitting model should be approximately normally distributed. To produce the probability plot we order the residuals and then calculate for each row the inverse of the cumulative density function for a normal distribution (Heckert et al., 2002). Effectively, we calculate for each of the ordered residuals the value that we would expect at this position for a normal distribution. These pairs are then visualized in a scatter plot. Systematic deviations of the points in the plot from the diagonal indicate that the empirical distribution does not match the theoretical distribution. Figure 5.2 shows the two probability plots corresponding to our constructed example, for the original and for the log-transformed target values. In this example, we easily see that the residuals for the original target values are not normally distributed while log transformation leads to an almost perfect match.

The QQ-plot is similar to the probability plot (Heckert et al., 2002), but instead of comparing an empirical distribution to a theoretical distribution we compare two empirical distributions. In the QQ-plot we simply plot the pairs of the percentiles of both distributions. This allows us to compare two samples of different size e.g., the empirical distributions of residuals on the training set and the testing set.

More detailed analysis of the model fit is possible through scatter plots of the residuals over each of the input variables. These *residual plots* allow to check for model bias, such as an effect that is not captured by the model or a systematic deviation of the model not supported by the data.

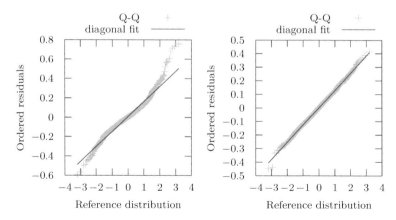

Figure 5.2: The probability plot for the residuals for the original target values (left) that are not normally distributed. The log-transformed target values (right) are normally distributed.

To demonstrate the residual plot we use "a simple function of ten variables" (Friedman et al., 1983; Friedman, 1991) shown in Equation (5.1).

$$f(x) = 10\sin(\pi x_1 x_2) + 20\left(x_3 - \frac{1}{2}\right)^2 + 10x_4 + 5x_5 \qquad (5.1)$$

$$Y = f(X) + \epsilon, \ X \sim_{iid} U(0,1)^{10}, \ \epsilon \sim_{iid} N(0, \sigma_{\text{noise}}), \ \sigma_{\text{noise}} = 0.21$$

This example function is easy to understand and analyze. It has four independent terms, two of which are linear in only one variable. The quadratic term depends only on x_3. The first term is nonlinear and an interaction of x_1 and x_2. We will use this function multiple times for demonstration in this chapter. For demonstrating residual plots we generate a dataset consisting of 100 observations of ten randomly distributed input variables in the unit cube, of which only five are relevant for the prediction of y. As in Equation (5.1) we add Gaussian noise $\epsilon_i = N(0, \sigma_{\text{noise}})$ so that the signal-to-noise ratio is high (23.4). The true underlying function accounts for 99.8% of variance of the response.

The residual plots for Equation (5.1) are shown in Figure 5.3. Each panel in the plot shows a scatter plot of the residuals of the model over one of the input variables. Each plot shows the zero as a reference and the mean and 95% confidence interval of a scatter plot smoothing model (LOESS) (Cleveland, 1979). To generate this plot we use the generating function Equation (5.1) and therefore no systematic deviations from zero should be seen in any of the plots. The smoothed value is close to zero almost everywhere, so we can safely assume that there is no remaining model bias. The plots for relevant variables are similar to the plots for the irrelevant variables ($x_6, \ldots, x_{10}$).

Figure 5.4 shows an example where model bias is apparent. To generate

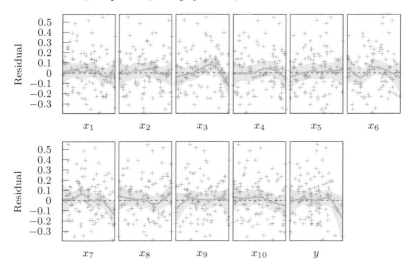

Figure 5.3: Residual plots for Equation (5.1). No systematic deviation of the model can be spotted in this example.

these plots we use an SR model found for the dataset instead of the generating function. To enforce model bias the variables x_1 and x_3 were purposefully not allowed. As a consequence the residual plots for this model show patterns of the residuals over x_1, x_3, and x_6 which means that there is information remaining in the data that is not yet captured by the model.

If such patterns occur in a residual plot, they should be scrutinized and the model should be extended to capture these patterns if reasonable. However, be careful distinguishing random noise from a signal. There is also the danger to manually overfit the model in a loop of model adjustment and visualization (Burnham and Anderson, 2003).

Closely related to the residual plot is the *partial dependence plot* (PDP) (Friedman, 2001). While the aim of the residual plot is to detect systematic deviations of the model from the data, in the PDP, the output of the model is visualized and its aim is to explain the model. The PDP allows to examine the predictions of the model over the full input space, in comparison to the residual plot which shows errors only for the observations in the dataset.

In the PDP, a separate panel for each input variable shows how the output of the model depends (partially) on that variable. It is important to note that the plot visualizes information about the distribution of the model output. For example to plot how the model response changes we assume that x_1 is independent from the other input variables and we take the expected value over the combined distribution of all other variables $x_2, \ldots x_d$.

$$\bar{f}(x_1) = E_{x_2,\ldots,x_d}[f(\boldsymbol{x})] = \int f(x_1, x_2, \ldots, x_d)\, p(x_2, \ldots, x_d)\, \mathrm{d}x_2, \ldots, \mathrm{d}x_d$$

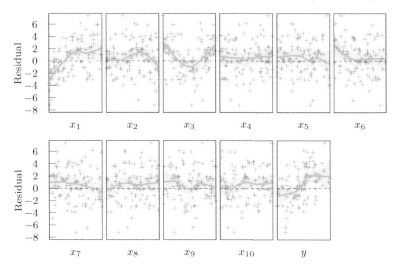

Figure 5.4: Residual plots for an SR model without using x_1 and x_3. Compared to Figure 5.3 patterns in the residuals over x_1, x_2, and x_6 are visible (note the different scale for the y-axis). Scatter plot smoothing helps to see the pattern.

Since the combined distribution of input variables is usually unknown, the empirical expected value is used as an approximation.

$$\bar{f}(x_1) = \frac{1}{N} \sum_{i=1}^{N} f(x_1, x_{i,2}, \ldots, x_{i,d})$$

We calculate the average repsonse over all observations of $x_2, \ldots, x_d$ in the dataset for each x_1. In the PDP we visualize this expected value $\bar{f}(x_1)$. Additionally to the average value we can calculate selected percentiles from the empirical distribution to visualize the variance of the model over all inputs for each x_1 value. The approximation used for the PDP is trivial to calculate while keeping the distribution of the inputs almost intact. However, the correlation between the selected variable for the plot and all other variables is broken because it is set explicitly and independently from the other values. This may lead to misleading results for datasets with correlated input variables. Figure 5.5 shows the PDP for the model Equation (5.1) for the dataset with 100 data points. The artifacts visible in the panels for x_1 and x_2 are caused by the small sample.

Closely related to the partial dependence plot is the *intersection plot*, where we do not attempt to approximate the marginal distribution but instead show the predictions of the model conditional on all input variables. As for the residual plot and the PDP each subplot concerns one of the input variables and shows how the prediction changes when only this variable is varied, assuming all other inputs are fixed. Correspondingly, each subplot in the intersection

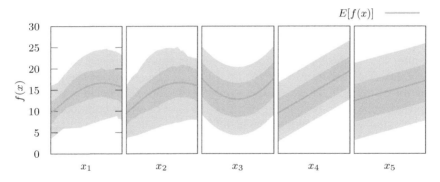

Figure 5.5: Partial dependence plot for Equation (5.1). Shaded areas show 50% and 90% probability intervals.

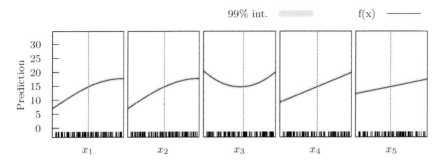

Figure 5.6: The intersection plot shows the response of the model over each of the inputs assuming all other variables are held fixed (marked by the dotted vertical lines).

plot depends not only on the analyzed variable but also on the values used for all other inputs. At the bottom of each sub-plot we show the distribution of values in the training set. This gives an indication for which input values the model is well supported by the data and for where the model is interpolating or extrapolating.

The intersection plot helps to detect problematic responses of the model which would not be apparent when only the responses for inputs in the training or validation sets are analyzed. The intersection plot shows the model response for a new sample of points convering the whole range of possible input values. Similarly to the residual plot and the PDP this visualization assumes that all input variables can be changed independently.

Figure 5.6 shows the intersection plot for Equation (5.1) with the response of the model and a 99% prediction interval (see Section 6.3). The intersection plot is mainly useful in interactive visualizations where the users can change each of the inputs in a kind of "what-if analysis". In this way, intersection plots allow users to explore how the model reacts to changes of input variables which can be especially interesting for models with interacting inputs.

It is important to keep in mind that the PDP and the intersection plot assume that each of the input variables is independent from the other input variables. This assumption is required to plot the output of the model over each input variable. However, this can be misleading in modelling tasks where input variables are correlated because the plot shows predictions for input vectors that would be impossible or unlikely to occur in reality. Visualization of the prediction uncertainty can be useful especially for input values that are far away from training points.

The visualization tools described in this section help to understand the input-output behaviour of the model but do not help to understand the internals of the model. In the next section we cover methods for in-depth explanation of SR models.

5.1.2 Explaining Models

A model's trustworthiness is highly dependent on our ability to explain its behaviour and understand how predictions are generated. Explaining models is a process of investigating and sometimes reverse engineering the internal structure and operating rules of a ML model for the purpose of better understanding the underlying relationships between variables. In the context of SR where models represent mathematical expressions, the process of explaining the model consists in understanding the flow of information through the model structure and explaining nonlinear transformations of input values and local interactions between the inputs.

A related problem towards this goal is estimating variable importance or broadly speaking, how much prediction models rely on input variables to generate accurate predictions. There are many different alternatives for how this can be achieved. Many techniques do not depend on a particular modelling method and can be used for SR as well as other regression methods. It is worthwhile to use several different ways to calculate variable importance and to compare the results.

In the following we distinguish between *model-specific* and *global variable importance* scoring methods. Model-specific importance quantifies the importance of variables in a given model while global importance refers to the importance over the whole modelling pipeline and does not require a validated model. *Permutation variable importance* perturbes information available in a variable by permuting the observed values and can be used for a given model or as a global measure. When using GP for SR, we may use the population of candidate models to quantify variable importance for example via *variable frequency.*

The choice for one or the other option often depends on the computational complexity. For deterministic algorithms, it is not necessary to distinguish between model-specific variable importance and global (algorithm-specific) performance because a (configured) algorithm deterministically only produces a single model. If the algorithm has hyperparameters, they may however affect the variable imporance score.

5.1.2.1 Variable Importance

An SR model represents a (nonlinear) transformation of the input values into an output representing the prediction. In many applications, such as in the medical field, it is critical to fully understand the calculation process of the model. At the very least, it is important to quantify how important each of the variables is for the overall prediction and – if we assume that the model is valid – also for the underlying system. The variable importance by itself can already give important insights about the modeled system.

Variable importance can be quantified by calculating how much of the variance of the target variable can be explained by each of the input variables, and is conceptually related to the partial dependence plot discussed above. A variable is important when it allows us to explain a large fraction of the variance of the target variable. Correspondingly, a variable is not important if it gives us no information about the target variable. In other words, if the target does not depend on an input variable then the variable is irrelevant. The concept can be extended to sets of variables to calculate the combined importance of multiple variables.

The calculation of variable importance is crucial in stepwise variable selection methods which are performed either in a forward or backward direction. Forward variable selection starts from a constant model and in each step adds the most important variable to the model. Backward selection starts from a model that uses all variables and in each step removes the least important variable.

Forward Variable Importance

For a given model $f(x_i))$ using only x_i as input, the variance explained by the model is

$$\text{VarExp}_{\text{forward}} = R^2(\boldsymbol{y}, f(\boldsymbol{x}_i)) = 1 - \frac{\text{var}(\boldsymbol{y} - f(\boldsymbol{x}_i))}{\text{var}(\boldsymbol{y})} \tag{5.2}$$

and is related to the coefficient of determination $R^2(\boldsymbol{y}, f(\boldsymbol{x}_i))$ when we assume that the model $f(\boldsymbol{x}_i)$ is linearly scaled such that the residuals $\boldsymbol{y} - f(\boldsymbol{x}_i)$ have zero mean and the same variance as the target values.

To calculate the variance explained by an input x_i we have to find an optimally fitting model $f(x_i)$. In forward variable selection one already has a model using a subset of the variables and calculates the variable explained when the model is extended with each of the remaining features. This allows to calculate the importance of sets of variables up to the full set of variables which leaves only the variance of the unexplainable noise term. For models where the effect of each input variable is additive, which requires that the inputs are independent, the variance explained by the full model is the sum of the VarExp values of each term.

Backward Variable Importance

We can also determine the importance of a variable as the relative decrease in explained variance when removing variable x_i from a given model or from the modelling pipeline completely

$$\text{VarExp}_{\text{backward}} = \frac{\text{var}(\boldsymbol{y} - f_{\backslash x_i}(\boldsymbol{X})) - \text{var}(\boldsymbol{y} - f(\boldsymbol{X}))}{\text{var}(\boldsymbol{y})} \qquad (5.3)$$

which is related to the R^2 reduction compared to the full model $R^2(\boldsymbol{y}, f(\boldsymbol{X})) - R^2(\boldsymbol{y}, f_{\backslash x_i}(\boldsymbol{X}))$. When the model $f(\boldsymbol{x})$ has been generated with SR, we have to remove references to x_i and replace them with coefficients if necessary to produce $f_{\backslash x_i}(\boldsymbol{x})$. Afterwards, all model coefficients have to be refit to the data to determine the importance of x_i. Alternatively, we run the modelling pipeline multiple times with and without x_i and determine the variable importance as the decrease in average R^2.

Potential Issues

The variable importance score is always conditional on the other input variables used for the model. This is important when an effect occurs because of an interaction of variables or when variables are correlated. A simple example helps to describe the effect.

First, consider the model $y = x_1 x_2$ with $x_1, x_2 \sim_{\text{iid}} U(-1, 1)$. For a set of observations from this system $\text{VarExp}_{\text{foreward}}(x_1)$ and $\text{VarExp}_{\text{foreward}}(x_2)$ will both be small because any model using only one of the two variables will not be able to explain the variance in observed y values. Conversely, if we start from the full model, both $\text{VarExp}_{\text{backward}}(x_1)$ and $\text{VarExp}_{\text{backward}}(x_2)$ will be high because removing either one of the variables will decrease the R^2 of the model to zero. For this simple example this is obvious. The effect is less obvious but just as important for larger models where interactions of variables occur. In such cases the importance of sets of variables should be considered additionally to individual variable importance. We will see the same effect again in the example below.

Moving forward, consider a the model $y = x_1 + x_2$ where x_1, x_2 are highly correlated. In this case the importance is high for both variables individually but when considered in a set (and removing one of the variables) the importance seems much smaller. In this case the importance score from removing a variable may underestimate the relevance of a variable in a system.

Colinearity (or, more generally, correlation spanning multiple variables) is especially problematic because it means that the importance of variables cannot be calculated separately and instead combinations of multiple variables must be considered. Shapley values are a game-theoretic concept and used to weight contributions of individual players within "coalitions". In regression modelling the concept is translated and used to assign contributions to individual variables within groups of variables used within a model (Lipovetsky and Conklin, 2001). The concept allows to determine variable importance over all combinations of variables, which leads to a better estimation of variable importance in the presence of multicolinearity in the data. Shapley values are used within Shapley additive explanations (SHAP) (Lundberg and Lee, 2017; Lundberg et al., 2020). Unfortunately, the computational effort to create models for all 2^d combinations of variables is often prohibitive. Shortcuts to reduce computation time are possible for partially separable or linear models. The method is especially useful for random forests or gradient boosted trees as it has been shown that these models allow a computational shortcut to make the calculation in polynomial time (Lundberg et al., 2020). Unfortunately, no similar shortcuts have been described for nonlinear regression models as produced by SR so far.

Global Variable Importance

The most general way to determine variable importance is to run the entire modelling pipeline twice. In the first run we include the variable and in the second run we exclude the variable. The importance of the variable can then be measured for example using a relative or absolute increase of the error measure to the likelihood.

This approach is completely independent of the algorithm used. However, it can also be costly because it my require many executions of the whole modelling pipeline – which may include hyperparameter optimization and cross-validation. If we are interested in the importance of variable combinations such as pairs or triples we require even more executions of the algorithm. Therefore, this way of determining variable importance globally can only be recommended for regression algorithms that are computationally inexpensive. In some special cases, for example linear regression, one can reuse a solution for more variables to calculate the solution for a reduced set of variables more efficiently (best subset regression, (Hastie et al., 2001; Miller, 2002)). For GP-based SR global variable importance is usually prohibitively costly.

Aggregate Variable Frequencies

In Section 4.8.2 we have briefly mentioned that in GP we can use the population of candidate models to estimate variable importance by counting the frequency of variables in the population. This gives us a more robust variable importance score as it does not depend on a single model. For instance if two important variables are highly correlated, some models use only the first while others use only the second. Over the whole population the number of references is similar for both variables, assuming the population is diverse enough. Therefore, we may get a better estimate for the importance of the two variables when we include all model candidates in our analysis.

The drawback of this method is that the number of variable references within models is only a crude surrogate for the importance. For example, within

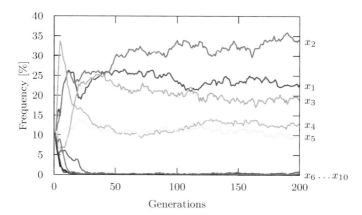

Figure 5.7: Variable frequencies over generations from one GP run for Equation (5.1). The irrelevant variables are quickly removed from the population while the more important variables are inherited by later generations and referenced more frequently.

a given model each variable may occur only once but their contributions to the prediction may be different. Similarly, a variable that occurs twice in a model does not necessarily have a stronger effect on the prediction. Variable frequency still works surprisingly well due to evolutionary selection pressure on the population of diverse models, which removes variables that are not important. At the same time, more important variables are copied more often because they occur in models with higher fitness which have a higher probability to be selected.

Figure 5.7 shows the variable frequencies from one GP run for the Friedman function (Equation (5.1)). In the initial population all variables occur with the same probability. After a few generations, the irrelevant variables are quickly removed from the population while the frequency of the most relevant ones quickly increases because they occur in the best models. The mean frequency over the whole run or over the last few generations can be used as a surrogate for variable importance.

Model-specific Variable Importance

Global variable importance and variable frequencies do not depend on a single model and therefore tend to be relatively robust. Additionally, bootstrapping or cross-validation can be used to further reduce the variance of the importance estimates.

However, once we have a validated well-fitting model we may also use this model or a small set of candidate models for the determination of variable importance instead of aggregating the expected importance value over a large set of models.

This model-specific importance score must be distinguished from the general variable importance score described above because a fixed model structure is assumed and $f_{\backslash x_i}(\boldsymbol{x})$ is a reduced version of the full model $f(\boldsymbol{x})$. This means we cannot freely fit the reduced model to maximize the explained variance.

For a given predictive model $M = f(\boldsymbol{x}, \hat{\boldsymbol{\theta}})$ with coefficients $\hat{\boldsymbol{\theta}}$ obtained from fitting the model to data we can determine the variable importance as a ratio

$$\text{Importance}(\text{M}, x_i) = \frac{\text{SSR}(M_{\backslash x_i})}{\text{SSR}(M)} \tag{5.4}$$

where $\text{SSR}(M)$ is the SSR of the original model, and $\text{SSR}(M_{\backslash x_i})$ is the SSR after removing variable x_i and refitting. The variables x_i are then ordered by the SSR ratio whereby important variables have a large SSR ratio. Alternatively, we may use the differences in the model selection criteria (ΔAIC, ΔBIC, and ΔDL) described in Section 5.2.1 if we want to take into account the reduction in model complexity as well. While the SSR ratio can be used only for least squares problems, the model selection criteria are more general and can be used with any likelihood. For the reduced model $M_{\text{reduced}} = f_{\text{reduced}}(\boldsymbol{x}_{\backslash x_i}, \hat{\boldsymbol{\theta}}_r)$ we have a new and potentially longer coefficient vector because we need to replace x_i by a new coefficient. The coefficient vector for the full model $\hat{\boldsymbol{\theta}}$ can be used as the starting point for fitting the $\hat{\boldsymbol{\theta}}_r$ of the reduced model.

This removal of variables merits a more detailed explanation because it is especially relevant to SR, where it is possible to include variables without a coefficient. To clarify how the variable is removed it is useful to study an example. Consider the model shown in Equation (5.5). For x_1 we may simply remove the first term to get a (re-parameterized) model Equation (5.6) whereby $\theta_{a,2}$ should be initialized to $\theta_{a,2} = \theta_3 + \text{mean}(\theta_1\, x_1)$.

When removing x_2 the term $\exp(\theta_2 x_2)$ becomes constant and should be integrated into $\theta_{b,2}$ using the initial value $\theta_{b,2} = \theta_3 + \text{mean}(\exp(\theta_2 x_2))$.

$$f(x, \boldsymbol{\theta}) = \theta_1 x_1 + \exp(\theta_2 x_2) + \theta_3 \tag{5.5}$$
$$f_a(x_2, \boldsymbol{\theta}_a) = \exp(\theta_{a,1} x_2) + \theta_{a,2} \tag{5.6}$$
$$f_b(x_1, \boldsymbol{\theta}_b) = \theta_{b,1} x_1 + \theta_{b,2} \tag{5.7}$$

This implies that the feature is important in the model if the reduced SR model is much worse after removing the variable and refitting coefficients. In contrast it is irrelevant if it can be removed without loss in accuracy.

Permutation Variable Importance

For an individual model or a small set of candidates we may quantify variable importance by calculating the increase in prediction error when permuting variable values. This approach was originally introduced for random forests and boosted regression trees. Breiman (2001) suggests permuting the values of each individual variable measuring the percent increase in the prediction error. The relative increase of prediction error is assimilated to a measure of variable importance, following the intuition that permuting the values of an important variable would cause a higher increase in prediction error than permuting the values of an unimportant variable. Breiman (2001) called this measure the *permutation variable importance score*. The procedure is related to the calculations required for the partial dependence plot described above and has the same advantages and disadvantages. It is easy to calculate but can produce misleading results when the dataset contains strongly correlated input variables, or when interactions of variables are relevant for the prediction. By permuting the values of a variable the correlation with the target variable (and other input variables) is broken and this can lead to misleading importance values.

Permutation variable importance scores can be aggregated for multiple candidate models to get a more robust score.

More recently, permutation importance has been generalized to other prediction models by Fisher et al. (2019) where it is called *model reliance*. Notably, the model reliance gives an indication of how much a model may rely on a variable while still predicting well.

Example

To demonstrate variable importance calculation methods, we again use the Friedman function introduced in Section 5.1.1.

$$f(x) = 10\sin(\pi x_1 x_2) + 20\left(x_3 - 0.5\right)^2 + 10x_4 + 5x_5 \qquad (5.8)$$

Using the known generating function and the distribution of input variables, we can establish reference values for the variable importance and use them to compare the variable importance quantification methods. The irrelevant variables $x_6, \ldots, x_{10}$, have zero importance by definition. The variance of $10\,x_4$ is $8\frac{1}{3}$, the variance of $5\,x_5$ is $2\frac{1}{12}$. The variance of the quadratic term is $2\frac{2}{9}$. The variance of the nonlinear first term is 11.188 (determined by Monte Carlo sampling). The variance of $f(x)$ is the sum of the variances of the terms and is 23.827. The ranking of terms by explained variance is correspondingly: $t_1(x_1, x_2)$ (47%), $t_3(x_4)$ (35%), $t_2(x_3)$ (9.3%), $t_4(x_5)$ (8.7%).

To calculate the importance of individual variables x_1 and x_2, which both occur in the first term, we have to find a function of only x_1 for which R^2 is maximal. The nonlinear contribution of the first term to Equation (5.8) is shown in Figure 5.8. The plot on the right shows multiple equally spaced intersections of the function and $m_1(x_1) = E_{x_2}(\sin(\pi x_1 x_2))$. The expression $y - m_1(x_1)$ has minimal variance and is therefore used to quantify the importance of x_1.

Using the definition of $m_1(x_1)$ we can calculate its variance which is close to 4.7. Therefore variance explained by x_1 (and for x_2 because of the symmetry) is approximately 20%. Finally, we know the order of the individual variables by their importance is: x_4 (35%), x_1, x_2 (20%), x_3 (9.3%), and x_5 (8.7%). The interaction of x_1 and x_2 in the first term means that the variance explained by both variables individually (as well as the sum) is smaller than the variance explained by both variables in combination.

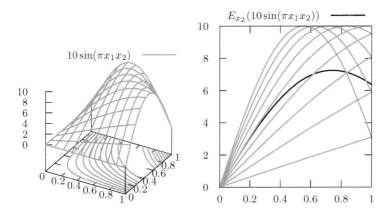

Figure 5.8: First term of Equation (5.8) (left), and expected value $E_{x_2}[10\sin(\pi x_1 x_2)]$ for x_2 uniformly distributed in the unit interval (right).

It is relatively easy to determine variable importance for this example with the methods introduced above despite the low number of observations, because the inputs are not correlated, observations are uniformly distributed over the whole input space, and the signal to noise ratio is high. Correspondingly all methods should produce similar importance scores. In actual modelling tasks this process is complicated by high dimensionality, a small number of observations relative to the number of input variables, low signal-to-noise ratio, and correlations between input variables.

Table 5.1 shows the results of the variable importance methods explained above. The row for *Global Importance* shows the ranking by explained variance when removing each variable completely from the modelling pipeline. The importance is calculated as the difference between the average R^2 of the models found with all variables and the average R^2 of the reduced models over 30 independent repetitions of GP. Here we use the R^2 values on testing partition. The algorithm configuration is optimized for each variable set using the best average test R^2 for different maximum tree sizes (10, 15, 20, 25, 30, 35 nodes) and number of generations.

The global importance is close to the reference ordering. It cannot distinguish well between variables that have similar importance. The irrelevant variables $x_6, \ldots, x_{10}$ all have an importance value close to zero and are not shown.

The row *Permutation Importance* shows the explained variance as the difference between test R^2 of the full model and R^2 when each variable is shuffled. The calculation is for the generating function Equation (5.1). The ranking produced by permutation importance is close to the optimal ordering. x_1 and x_2 have very similar scores and their order is swapped relative to the reference. The importance values are overestimated as a consequence of the permutation scheme.

Table 5.1: Comparison of different variable importance scores for Equation (5.1). The different methods produce similar orderings for the variable importances.

Method	x_4	x_1	x_2	x_3	x_5
Reference	35%	20%	20%	9.3%	8.7%
	(1)	(2)	(3)	(4)	(5)
Global importance	54%	28%	35%	12%	9.9%
	(1)	(3)	(2)	(4)	(5)
Permutation importance	61%	40%	44%	14%	11%
	(1)	(3)	(2)	(4)	(5)
Model importance	30%	29%	19%	8.7%	6.8%
	(1)	(2)	(3)	(4)	(5)
Variable frequency	13%	24%	30%	20%	11%
	(4)	(2)	(1)	(3)	(5)

The row *Model Importance* shows the results when removing each variable from the optimal model and refitting the coefficients. The variable importance scores are very close to the reference values and lead to the same ordering. However, the variance from the limited sample of 100 observations causes a much higher importance score for x_1 compared to x_2 even though they should have the same importance.

The last row shows the average variable frequencies from a single GP run. The importance values are the average frequencies over all generations from the run shown in Figure 5.7. The variable frequency results for a single GP run differ the most from the other methods which can be explained by the fact that variable frequencies are only a surrogate for variable importance. Additionally, variable frequencies vary strongly from one GP run to the next. The variance can be reduced by executing multiple GP runs and taking the average. The irrelevant variables all have average frequencies below 1% and are not shown.

5.1.2.2   Subexpression Importance

We use the more general term *subtree importance* or *subexpression importance* to describe not only a single variable's importance score but an entire subexpression's importance within an SR model. The calculation of subexpression importance is similar to the calculation of model-specific variable importance. The only difference is that a whole subexpression is replaced by a coefficient instead of a variable. This serves as a reliable indicator of whether a subexpression has a large enough influence on the model response or can otherwise be removed to reduce complexity and improve interpretability. We discuss model simplification using subexpression importance in Section 5.3.2.

5.1.3 Model Interpretability

We often deal with complex systems whose behaviour cannot be fully formalized using first principles. It is important in such cases to obtain models that capture correlations in a way that they can be analyzed and interpreted by users. We recommend to use SR especially for modelling tasks where the objective is primarily to generate insights and understand the relationships between variables. This is in contrast to tasks where we primarily need an accurate predictive model and interpretability is a secondary concern.

Interpretability in ML is difficult to quantify and thus cannot be explicitly promoted during model development. The practical approach is to optimize for properties that can impact interpretability such as model size or related measures of complexity and balance these and predictive accuracy. Interpretability – meaning the capability to assign physical meaning to components of the model – needs to be defined on a per-application basis. To quantify this property, a SR model is usually assessed by experts using criteria derived from domain knowledge, such as:

- *Structural constraints.* These constraints restrict the hypothesis space to certain structural patterns and are typically applied when the structure of the model is partially known or shall be enforced for interpretability.

- *Behavioural constraints.* This category of constraints refers to the behaviour or shape of the learned function, in terms of monotonicity, smoothness, properties of its derivatives with respect to the independent variables, and similar properties.

- *Physical constraints.* More specific structural or behavioural constraints based on physical plausibility, for example: a mathematical function that is dimensionally consistent, expressing a plausible relationship between physical quantities, or behaving in a way consistent with existing physical laws, theory, or experimental data.

- *Complexity constraints.* Complexity reduction is an important objective towards interpretability. Many complexity measures for SR models exist in the literature. Typically, these measures refer to the size and shape of the model, the presence of nonlinear functions, interactions between variables, or the nesting level of nonlinear functions.

The constraints listed above can be seen as an initial set of requirements for interpretability. The end result and whether or not it brings additional insight to the table, depends on the overall success of the knowledge discovery pipeline, the quality of the raw data and the data preprocessing step, and the amount of prior knowledge and information available. These constraints should be checked for candidate models for model validation and selection. Additionally, we may use algorithmic techniques to ensure that the produced models fulfill such constraints. In particular, structural constraints can help identify models composed of multiple interpretable terms. These techniques are discussed in more detail in Section 6.1.

5.2   Model Selection

In the following, we use the term *model selection* to describe the process of choosing a single model (or a small number of models) from a large set of candidate SR models that have been generated using GP. We note that this is different from model selection during training or algorithm tuning or implicit selection performed for example by GP during evolution.

For algorithmic model selection we may use the same criteria that are used in manual model selection as long as they are quantifiable. Manual model selection may however also include subjective decisions based on qualitative assessment, for instance based on visualizations, background knowledge, personal preferences, or restrictions imposed by downstream processes where the models are used. The main objective should be to find a model that is plausible and useful.

Time spent for careful model selection is well-invested because this final step in the modelling pipeline can determine the success of the entire modelling task. A mistake in model selection can negate all the efforts spent on data collection and preparation, algorithm selection, or hyperparameter tuning.

A necessary prerequisite for model selection is that we have a larger set of maybe ten to one hundred models that are all well-fit and validated using the model validation techniques described above. We cannot reject any of the remaining models directly.

We recommend selecting a handful of different models when they all seem equally plausible and when it is possible in the application. This ensembling of models reduces the risk of incorrect model selection and stabilizes the variance of model predictions. However, downstream processes sometimes require to select a single model. In this case, careful model selection is required.

Model selection for regression models is covered well in many good books (see the references given at the end of this chapter). We will focus on the peculiarities of SR especially in combination with GP. The basics of model selection in particular underfitting and overfitting as well as cross-validation are summarized in Section 2.5.

5.2.1   Criteria for Model Selection

Given a set of alternative SR models e.g., from multiple GP runs, we can rank the models using several criteria such as the error on a hold-out set, criteria motivated by information-theory such as Akaike's information criterion, the Bayesian information criterion, minimum description length, or Bayes factors based on the Bayesian model evidence. In-depth discussion of model selection and corresponding criteria can be found for instance in Hastie et al. (2001), Burnham and Anderson (2003), and Gelman et al. (2013).

5.2.2 Hold-out Set for Validation

We have already briefly discussed the concept of hold-out sets for validation or testing. If we are in a data-rich situation and have a hold-out set that has not been used at all in the previous training steps then we can use it for model selection. The hold-out set should be large enough to make sure that the prediction error for the hold-out set is a good estimator for the expected prediction error for new data.

Using a hold-out set for model selection is the best option when sufficient data are available (Hastie et al., 2001). Unfortunately, we often do not have the luxury of a large enough dataset from which we can remove data for the hold-out set and still have enough data remaining for fitting. Additionally, the hold-out set can only be used once in the modelling pipeline. Strictly speaking, once the models have been evaluated on the hold-out set and we have made decisions based on the results, the hold-out set must not be used again. This means that if we expect that we need multiple modelling iterations including model selection, we would need to store away several hold-out sets at the beginning of the project. Repeated evaluation and optimization against the hold-out set can lead to overfitting to the hold-out set with the same negative implications as overfitting to the training set.

5.2.3 Cross-validation

k-fold cross-validation (CV) (Stone, 1977) uses the available data more efficiently by partitioning the data into k evenly sized folds. Training is repeated k times whereby each fold is used as a hold-out set once. Five-fold or ten-fold CV provides a good tradeoff between computational effort and estimation accuracy of the generalization error (Hastie et al., 2001). The cross-validation error (CV error) is the average MSE on the hold-out sets (Equation (5.9)).

CV is attractive because it uses the available data efficiently. The CV score is a good estimator of the generalization error, because we virtually use the whole dataset as a hold-out set. However, the computational effort of CV is higher, because multiple runs of the algorithm have to be performed. It is important to note that CV does not give a score for an individual model but a score for the algorithm and its parameters. Therefore, it can be used to optimize hyperparameters but cannot be used to select one out of multiple fitted SR models.

For SR we recommend CV to choose the algorithm parameterization which generates the most parsimonious models and whose CV error is no more than one standard error of the CV error of the best parameterization (Breiman et al., 1984). For example, Hastie et al. (2001) state: *"Often a 'one standard error rule' is used with cross-validation, in which we choose the most parsimonious model whose error is no more than one standard error above the error of the best model."*

More formally, given the MSE values for k folds for algorithm parameters η the CV score is

$$\text{CV}(\eta) = \frac{1}{k} \sum_{i=1}^{k} \text{MSE}(\eta)_i \tag{5.9}$$

the standard deviation is

$$\text{SD}(\eta) = \sqrt{\text{var}\left(\text{MSE}_1(\eta) \ldots \text{MSE}_k(\eta)\right)} \tag{5.10}$$

and the standard error of the CV score $\text{SE}(\eta) = \text{SD}(\eta)/\sqrt{k}$.

If the parameterization $\eta^\star \in \{\eta_1, \ldots, \eta_m\}$ is the one with the best CV score $\text{CV}(\eta^\star) = \min_{\{\eta \in \eta_1, \ldots, \eta_m\}} \text{CV}(\eta)$. The one standard error rule implies to instead select the parameterization η with

$$\text{CV}(\eta) \leq \text{CV}(\eta^\star) + \text{SE}(\eta^\star) \tag{5.11}$$

which produces the least complex models.

CV as originally described in the literature uses a random assignment of observations to partitions (shuffling). This is of course only allowed when the assumption of independent observations is correct. If subsequent rows in the dataset are not independent, for instance because they are measurements from a time series, or because the data stem from an experiment with grouped observations, then we cannot use a random assignment to folds. The reason is that random assignment causes the training and test sets to contain similar observations and the CV score will therefore be too optimistic. Alternative model selection schemes similar to CV are possible in such situations. The general idea is always to make sure that the distribution of observations in the hold-out sets is similar to future observations and the similarity of the training set to the hold-out set should be equal to the similarity of the training set to future data.

5.2.4   Akaike's Information Criterion

The Akaike information criterion (AIC) (Akaike, 1974; Burnham and Anderson, 2003) is a criterion for model selection that combines two terms to balance between model error and complexity. The first term expresses the fit on the training data and the second term can be interpreted as an approximation for optimism of the training error. It therefore does not require a hold-out set and can be used if data are scarce. This property is shared with BIC and DL described in the next sections. The second term depends only on the number of fitting parameters. The model which minimizes AIC has a high likelihood and few fitting parameters.

For the common assumption of independent and identically normally distributed errors for the model function $f(x, \boldsymbol{\theta})$ and $D = (\boldsymbol{X}, \boldsymbol{y})$ with n observations, the log of the likelihood function is

$$\log \mathscr{L}(\boldsymbol{\theta}, \sigma_{\text{err}}) = -\frac{n}{2}\log(2\pi\sigma_{\text{err}}^2) - \frac{1}{2\sigma_{\text{err}}^2}\sum_{i=1}^{n}(y_i - f(\boldsymbol{x}_i, \boldsymbol{\theta}))^2 \qquad (5.12)$$

Here, we have explicitly mentioned the parameter for the standard deviation of noise σ_{err}. In some applications, the standard error of measurements is known at least approximately. Otherwise, it can be estimated using a low-bias model (Hastie et al., 2001). For model selection, we use the same σ_{err} for all models. Alternatively, σ_{err} can be included in the vector of fitting parameters, in which case it must also be counted in the number of parameters k.

The AIC for a given model with maximum likelihood parameter estimate $\boldsymbol{\theta}$ is

$$\text{AIC} = -2\log \mathscr{L}(\boldsymbol{\theta}) + 2k \qquad (5.13)$$

where k is the length of $\boldsymbol{\theta}$.

For model selection we evaluate Equation (5.13) of each model on *the same dataset* and select the models with the smallest AIC values. The AIC value depends on n, the number of observations in the dataset. Therefore, we cannot establish thresholds for the AIC value as it can be large or small depending on n. Relevant is the *difference between* AIC values of two different models. We use the model with minimal AIC_1 as a reference and calculate the differences $\Delta_i = \text{AIC}_i - \text{AIC}_1$ for the i-th model. We can use a rule-of-thumb for model selection based on Δ_i (Burnham and Anderson, 2003): *"models with $\Delta_i \leq 2$ have substantial support (evidence), those in which $4 \leq \Delta_i \leq 7$ have considerably less support, and models having $\Delta_i > 10$ have essentially no support"*. This means that we may include models with $\Delta_i \leq 2$ in the reduced model set and we should certainly remove models with $\Delta_i > 10$. This is regardless whether $\text{AIC}_1 = 300$ and $\text{AIC}_i = 310$, or $\text{AIC}_1 = 300000$ and $\text{AIC}_i = 300010$ because only the difference between models matters. Δ_i can also be used to calculate *Akaike weights* for weighing models in an ensemble (Burnham and Anderson, 2003).

AIC is attractive for model selection because it can be used to compare different models, including models with different likelihood functions, as long as the dataset is unchanged (Burnham and Anderson, 2003). AIC is not limited to linear models. It is only required that the model has been fit via maximum likelihood estimation (MLE) (Burnham and Anderson, 2003). Least squares, as used frequently in SR, is equivalent to MLE with the assumption of independent and identically distributed Gaussian errors.

AIC approaches the generalization error in the limit of infinite data. For smaller datasets, say when $n < 40\,k$, AIC is too optimistic. "Optimism" in this context is taken in the sense that the likelihood overestimates the expected likelihood and the correction term used in AIC does not penalize this strongly enough for small datasets. Therefore, the adjusted version AIC_c is recommended instead (Burnham and Anderson, 2003).

$$\text{AIC}_c = \text{AIC} + \frac{2k(k+1)}{n-k-1} \qquad (5.14)$$

5.2.5 Bayesian Information Criterion

The *Bayesian information criterion* (BIC) is very similar to AIC but represents a more conservative tradeoff between likelihood and the number of parameters. It is an approximation of the Bayesian evidence whereby it approaches the evidence with increasing number of observations.

$$\text{BIC} = -2\log\mathscr{L}(\boldsymbol{\theta}) + k\log n \qquad (5.15)$$

Adding a parameter has a larger weight in the BIC than in the AIC for almost all practical datasets, and must be offset by a larger relative improvement of the likelihood compared to AIC. This has the consequence that BIC is more conservative in model selection. Compared to BIC, AIC has a higher chance to include smaller effects into the model (Burnham and Anderson, 2003).

AIC and BIC do not account for the complexity of the operators and functions used in the model as only the number of parameters is relevant for the calculation of the score. A linear model with five parameters is penalized equally to a model which includes many nested nonlinear functions as long as the number of parameters is the same. An alternative is to calculate the effective degrees of freedom as the trace of the Hessian of the parametric model and use it instead of the number of parameters (Hastie et al., 2001). Handley and Lemos (2019) suggest to use the *Bayesian model dimensionality* as an estimate for the effective number of parameters.

5.2.6 Minimum Description Length Principle

The *minimum description length (MDL)* principle recommends taking the length or the structural complexity of the model into account for model selection (Rissanen, 1978; Grünwald, 2007). Arguably, MDL is therefore better suited for SR than AIC or BIC, which only account for the number of model parameters.

The MDL principle states that we should select the model that allows the shortest encoded (compressed) combined representation of the model and the data. MDL literature uses the term *code* to refer to the encoded representation of the data and the model, and the central concern is the minimization of *code length*.

Predictive models can be used to compress data because they capture patterns in the data. If we assume a prediction model is available, we only have to transmit the prediction errors to allow restoring the original data. As the prediction errors of a well-fit model are are small, this requires fewer bits than transmitting the original data. Additionally, to the encoded prediction errors, the model itself must be transmitted to allow the receiver to fully restore the original data. Following the MDL principle, we should select the model with minimal code or description length $L(D) = L(M) + L(D|M)$, which is the sum of the model code length $L(M)$, and the code length of data compressed with the model $L(D|M)$.[1] $L(M)$ depends on the complexity of the model. Put simply, a model that requires more parameters requires a longer code than a model with fewer parameters. In SR, the code length is not only determined by the parameters but also by the length of the expression. The code length increases with the number of terms, functions, or operators that occur in the model. $L(D|M)$ depends on the goodness-of-fit of M. A model with on average smaller prediction errors allows a better compression of the data and therefore a shorter code. The two terms used in the description length again show the typical tradeoff in model selection; we have to chose a model that fits the data well, but at the same time has minimal complexity.

Description length is attractive for model selection because the code length of the model and the encoded data are both expressed in units of information (typically bits), which means that a weighting parameter is not required. However, the code length for the model has to be calculated carefully to accurately approximate the information content of the model. As a crude approximation, we could simply use the length of the expression multiplied by the code length required for each symbol. With this approximation, an expression of length w with u different symbols can be encoded using $w \log_2 u$ bits. This would however neglect the number of bits required for transmitting the coefficients of the model.

A better formula for the code length of a SR model is given in Equation (5.16) taken from Bartlett et al. (2023b).

$$L(D) = L(D|M) + L(M)$$

$$= -\log \mathscr{L}(\hat{\boldsymbol{\theta}}) + w \log u + \sum_j \log c_j + k \log 2 + \sum_i \log \frac{|\hat{\theta}_i|}{\Delta_i} \qquad (5.16)$$

This formula takes the constants (c_j) and coefficients θ_i occurring in the SR model into account for the calculation of $L(M)$. The code length required for each constant depends on its size and is $\log c_j$. The maximum likelihood estimate of the p parameters is $\hat{\boldsymbol{\theta}}$. The code length required for each coefficient $\hat{\theta}_i$ depends on the precision Δ_i^{-1} that is required for each parameter. The term $p \log 2$ is necessary to account for the sign of parameters. $L(D|M)$ can be approximated through the negative log-likelihood (Shannon-Fano theorem) (Cover and Thomas, 2005) evaluated at $\hat{\boldsymbol{\theta}}$. It is convenient to consistently use the natural logarithm in the formula and therefore the result is the approximate code length in nats.

The required precision Δ^{-1} can be estimated via the observed Fisher

---

[1] Note that here L is the code length not a likelihood.

information matrix I (Wallace and Freeman, 1987), which is related to the standard error of $\hat{\boldsymbol{\theta}}$.

$$(I(\boldsymbol{\theta}))_{i,j} = -\frac{\partial^2}{\partial\theta_i\partial\theta_j}\log\mathscr{L}(\boldsymbol{\theta}) \tag{5.17}$$

A large Fisher information value means that the likelihood has a strongly curved peak and therefore the parameter estimation is precise (low standard error). Using I to estimate the precision Δ^{-1} in Equation (5.16) leads to the approximation Equation (5.18) (Wallace and Freeman, 1987; Bartlett et al., 2023b) which only requires the diagonal of I and has no parameters that require tuning.

$$L(D) = -\log\mathscr{L}(\hat{\boldsymbol{\theta}}) + w\log u + \sum_j \log c_j - \frac{k}{2}\log 3 + \sum_i^k \left(\frac{1}{2}\log I_{ii}(\hat{\boldsymbol{\theta}}) + \log|\hat{\theta}_i|\right) \tag{5.18}$$

In this approach the precision of the parameter estimates is used to discretize each value individually for the calculation of the code length for each parameter. This can be inaccurate for highly correlated parameters. Bartlett et al. (2023a) later extended the approach to discretize the parameters jointly using the full Fisher matrix instead of only the diagonal. In the same work different priors for encoding parameters and the relation between description length and Bayesian model selection are discussed.

AIC and BIC are easy to calculate for SR models because we only require the likelihood of data given the model, the number of training observations n, and the number of parameters k of the model as well as an approximation of the noise variance σ_{err}^2. For calculating L(D) we additionally require the (approximate) observed Fisher information matrix which can be approximated using numeric differences or via automatic differentiation (see Section 6.2).

In summary, selection based on information criteria and description length can be recommended in cases when we do not have enough data for a separate validation set, or when CV is not possible because of computational limitations.

When used in combination with GP, AIC, BIC, and DL have to be applied carefully as those criteria do not account for the optimism from evaluating many different model structures and can therefore be misleading. Trying a large number of model structures and filtering by training error – which is exactly what GP is doing – may increase the probability of overfitting and would require another correction term to account for the number of model structures examined.

5.2.7 Comparison of Model Selection Criteria

To demonstrate the different criteria for model selection we again use the Friedman function (see Equation (5.1)). We generate 21 different models with tree-based GP using seven different limits for the tree length

Table 5.2: Model selection criteria for 21 models for the Friedman function ranked by test error. The top-5 ranked models for each criterion are marked. The models with best training error (RMSE$_{tr}$) are too complex and overfit. The top five models selected by AICc and BIC are the same in this example, only the ordering differs. BIC correctly identifies the top three models. For DL, the top five models are also the five models with best test error.

Rank	k	RMSE$_{te}$	RMSE$_{tr}$		AICc		BIC		DL	
1	11	0.227	0.189		−19.7	3	8.0	1	125.8	2
2	12	0.238	0.187		−18.9	4	10.8	2	142.8	4
3	14	0.248	0.178		−20.7	2	12.7	3	147.9	5
4	8	0.268	0.242		24.7		46.2		113.3	1
5	9	0.269	0.243		28.4		52.0		141.8	3
6	22	0.279	0.165		−6.2		39.2		249.1	
7	22	0.288	0.143		−21.6	1	23.8	4	219.9	
8	22	0.299	0.157		−11.8	5	33.6	5	230.5	
9	33	0.303	0.159		33.5		85.4		452.4	
10	35	0.325	0.102	1	9.5		61.0		472.5	
11	32	0.327	0.136	4	13.4		65.4		406.9	
12	39	0.328	0.116	3	37.6		86.2		450.1	
13	31	0.337	0.105	2	−7.8		44.1		349.5	
14	11	0.414	0.404		270.2		297.9		282.2	
15	38	0.444	0.140	5	46.4		96.0		553.3	
16	11	0.475	0.430		319.3		346.9		323.6	
17	9	0.555	0.550		580.8		604.4		427.0	
18	9	0.878	0.605		725.1		748.6		485.1	
19	7	1.098	1.046		2368.2		2387.5		1280.5	
20	7	1.631	1.378		4195.8		4215.0		2178.5	
21	7	1.631	1.378		4195.8		4215.0		2178.5	

$(15, 20, 25, 30, 50, 75, 100)$ and three random restarts. To compare the selection criteria we calculate the test error for all models using a million random samples from the same distribution. Table 5.2 shows the models ordered by increasing test RMSE with the number of parameters k, the training RMSE, AICc, BIC, and DL. For this dataset we know that $\sigma_{err} \approx 0.21$ and use this value for the calculation of the Gaussian likelihood for AICc, BIC, and DL. Table 5.2 shows values of model selection criteria for the best models ordered by test error. In summary, BIC works best in this simple example.

In comparison, Figure 5.9 shows the 10-fold cross-validation results for GP with tree sizes limit $\in (3, 5, 7, 10, 15, 20, 25, 30, 35, 45, 55)$. The plot shows the CV score over the ten runs with its standard error. For informational purposes, to see how well the CV score correlates with the test error, the plot also shows the MSE on the test set (with one million rows) for each of the ten models as well as their average. This information would not be known when we run

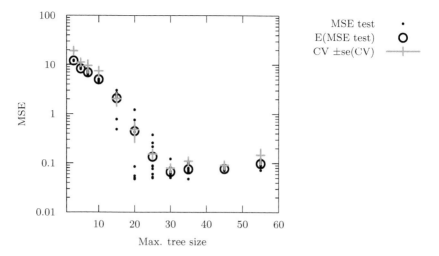

Figure 5.9: Cross-validation scores for the Friedman example. The CV score and the mean of test MSE values are correlated, which demonstrates the accuracy of CV for this example. The best size limit selected by CV is 30 nodes.

cross-validation. In this example, the CV score correlates well with the test error. Based on the CV scores we would select 30 nodes for the maximum tree size, which is also the configuration with the best test error. These models contain approximately ten to 15 parameters which aligns well with the results in Table 5.2.

5.3   Model Simplification

Once we have generated an SR model, we have to decide whether the model can be used as is or should be simplified. This can happen for example when using GP with an overly large limit for the tree size.

Structural transformations are primarily concerned with reducing the model to a simpler form, easier to evaluate and to interpret. We distinguish two main types of structural transformations:

1. Algebraic simplification of the model structure: this step involves using mathematical identities where applicable and folding of constants. This is necessary in GP-based SR because GP promotes structures that are easily extensible. The expressions produced as a result are therefore often unnecessarily complex and may often be simplified algebraically.

2. Pruning of the model structure: this step involves the removal of parts of the expression.

These structural transformations are typically useful because GP requires diversity of structures for evolvability and promotes bloated individuals that are more robust against deleterious evolutionary operations. Many GP systems tend to produce overparameterized models which pose problems for example when trying to fit and interpreting parameters. Therefore, solution candidates produced by GP can often be simplified/pruned while incurring only a small quality loss. de Franca and Kronberger (2023) have described a simplification method especially for GP-based SR which uses equality graphs and equality saturation to reliably remove redundant parameters.

Pruning is a specific case of model selection because we have to decide which subtrees can be removed to reduce the complexity without increasing prediction error too much. Thereby we implicitly have to consider a set of simplified versions of a model.

In the following sections, we first describe analysis and comparison of nested models, where coefficients are set to zero, and then extend the idea for the calculation of subexpression importance for pruning.

5.3.1 Nested Models

One way to simplify models is to remove coefficients and associated subexpressions by setting them to zero. Starting from a full model, we can generate a set of nested models, where each one is the result of removing one coefficient. The nested models can then be refit to the dataset. If the decrease in goodness-of-fit is not significant, we may accept the simplified model and repeat the procedure to find the next parameter that may be removed. This greedy, iterative approach is computationally cheap. Nested models may also be generated by setting multiple coefficients to zero at the same time, but the effort quickly grows with the size of the combinations.

For linear models and least squares, the exact likelihood ratio test can be used to determine if the reduction of variance explained is significant (Draper and Smith, 1998). It can also be used for nonlinear models but only as an approximation (Bates and Watts, 1988). The likelihood ratio test is applicable only for nested models, i.e., models that can be generated from the full model by setting coefficients to zero (Pawitan, 2001). The AIC and BIC are more generally applicable and may also be used here.

For example, for the Friedman dataset we found the expression

$$y \approx \theta_1 \sin(\theta_2 x_2 \theta_3 x_1) + \exp(\exp(\theta_4 + \cos(\theta_5 x_3)) + \cos(\theta_6 x_3)) + \theta_7 x_4 + \theta_8 x_5 + \theta_9 \tag{5.19}$$

with nine parameters $\boldsymbol{\theta} = (10.09, 7.56, 0.42, -1.62, -6.02, -6.38, 9.97, 5.15, -0.24)$ using GP. Simplifying the expression, removing unnecessary parameters

Table 5.3: Relative likelihood analysis for nested models of Equation (5.20).

	Δk	ΔSSR	F-ratio	p-value	ΔAICc	ΔBIC	ΔDL
θ_1	2	596.4	4559.5	$< 10^{-8}$	9536.4	9534.4	4734.8
θ_2	2	596.4	4559.5	$< 10^{-8}$	9536.4	9534.4	4734.8
θ_3	2	18.07	138.11	$< 10^{-8}$	283.2	281.2	116.0
θ_4	1	0.047	0.711	0.4039	-2.2	-3.2	-18.5
θ_5	1	0.129	1.9774	0.1670	-0.9	-1.8	-7.2
θ_6	1	439.6	6721.9	$< 10^{-8}$	7031.0	7030.0	3501.7
θ_7	1	81.26	1242.4	$< 10^{-8}$	1291.1	1296.2	632.4
θ_8	1	0.544	8.3244	6.15×10^{-3}	5.7	4.8	-5.1

and merging θ_2, θ_3 the new expression with eight parameters is

$$y \approx \theta_1 \sin(\theta_2 x_1 x_2) + \exp(\theta_3 \exp(\cos(\theta_4 x_3)) + \cos(\theta_5 x_3)) + \theta_6 x_4 + \theta_7 x_5 + \theta_8$$
$$(5.20)$$

with $\boldsymbol{\theta} = (10.09, 3.14, 0.2, -6.02, -6.38, 9.97, 5.15, -0.24)$. The model has MSE = 0.0578, AICc = 22, BIC = 34.72, DL = 109. Re-fitting the parameters to the dataset using nonlinear least squares we get the optimized parameter vector $\boldsymbol{\theta} = (10.1, 3.16, 0.207, -6.04, -6.42, 9.99, 5.27, -0.332)$ and MSE = 0.0549, AICc = 19.7, BIC = 32.4, DL = 108.5 for $\sigma_{\mathrm{err}} = 0.25$ on the training set spanning 50 rows. Fitting all nested models where one of the coefficients is set to zero we get the results shown in Table 5.3.

Following Bates and Watts (1988), we calculate the increase in the sum of squared residuals (SSR) relative to the number of removed coefficients $s_{\mathrm{extra}}^2 = \frac{\Delta SSR}{\Delta k}$ and the F-ratio for the relative likelihood test:

$$\text{F-ratio} = \frac{s_{\mathrm{extra}}^2}{s_{\mathrm{full}}^2} \qquad (5.21)$$

where $s_{\mathrm{full}}^2 = \frac{SSR_{\mathrm{full}}}{n-k}$ is the estimated noise variance for the n residuals of the full model with $n - k$ degrees of freedom. Using the F-distribution with degrees of freedom Δk and $n - k$ we can find the p-value for the F-ratio. In this example we find high p-values for θ_4 and θ_5 which implies that we should consider setting one of the parameters to zero.

We may also use ΔAICc, ΔBIC, and ΔDL instead of the likelihood ratio to check whether a nested model should be prefered. The last three columns show the difference in AICc, BIC, and DL (for $\sigma_{\mathrm{err}} = 0.25$). All three criteria improve slightly when setting either θ_4 or θ_5 to zero. Since $\theta_4 = 0$ is most strongly indicated to be zero by all three criteria we first try the simplified model

$$y \approx \theta_1 \sin(\theta_2 x_1 x_2) + \exp(\theta_3 + \cos(\theta_4 x_3)) + \theta_5 x_4 + \theta_6 x_5 + \theta_7 \qquad (5.22)$$

with $\boldsymbol{\theta} = (10.1, 3.16, 0.598, -6.22, 10.1, 5.15, -0.745)$. This model has 1.7% worse SSR but better AICc, BIC, and DL scores (MSE = 0.0559, AICc =

Table 5.4: Relative likelihood analysis for nested models of the simplified model (Equation (5.22)).

	Δn	ΔSSR	F-ratio	p-value	ΔAICc	ΔBIC	ΔDL
θ_1	2	604.6	4653.7	$< 10^{-8}$	9668.5	9666.2	4803.4
θ_2	2	604.6	4653.7	$< 10^{-8}$	9668.5	9666.2	4803.4
θ_3	1	18.02	277.38	$< 10^{-8}$	285.4	284.4	135.0
θ_4	2	93.85	722.35	$< 10^{-8}$	1496	1493.8	718.6
θ_5	1	450.6	6936.2	$< 10^{-8}$	7206.5	7205.4	3591.1
θ_6	1	77.93	1199.6	$< 10^{-8}$	1244	1242.9	608.9
θ_7	1	2.305	35.478	4.23×10^{-7}	34	33	11.3

17.5, BIC $= 29.3$, DL $= 89.5$). The results for nested models of the simplified model in Table 5.4 indicate that another step of simplification would decrease the likelihood of the model severely and we accept the simplified model.

Comparing the model with Equation (5.1) we see that the model is close to the generating function but failed to identify the correlation with x_3 correctly. Instead, the term $(x_3 - 0.5)^2$ is approximated via the cosine function.

5.3.2 Removal of Subexpressions

In Section 5.1.2.1 we have discussed model-specific variable importance where the prediction error of the full model is compared with the prediction errors of reduced models when a variable is removed. If we find that a variable can be removed with only a small decrease in the goodness-of-fit, the parsimony principle suggests to select the smaller model, as it is more likely to generalize to new data. The process may be repeated until no variable can be removed without an unacceptable loss in accuracy.

The same principle can be applied to subexpressions. If a subexpression has almost no effect on the response we should remove the subexpression to simplify the model. Because we analyze the effect of removing whole subexpressions, this process may uncover sets of variables that can only be removed in combination.

To calculate subexpression importance we follow the same process as described in Section 5.1.2.1 for the calculation of model-specific variable importance. We represent the expression as a tree and iterate the tree in post-order (bottom-up). For each of the nodes in the tree we create a copy of the whole tree but replace the node with a coefficient. This removes the whole subtree rooted at the replaced node from the model. After simplification and merging redundant coefficients, we refit the new expression and calculate the SSR and likelihood of the reduced expression. The importance of the subexpression is then determined via the SSR or likelihood ratio or via ΔAICc, ΔBIC, or ΔDL.

Table 5.5 shows the SSR ratio and increase in AICc, BIC, and DL for some subexpressions of Equation (5.22). From the values it is clear that no

Table 5.5: Importance of subexpressions in Equation (5.22).

Subexpression	SSR factor	ΔAICc	ΔBIC	ΔDL
$\theta_1 \sin(\theta_2 x_1 x_2)$	217	9669	9666	4803
$\theta_5 x_4$	162	7206	7205	3591
x_1	108	4779	4779	2385
x_2	108	4779	4779	2385
$\cos(\theta_4 x_3)$	34.6	1496	1494	719
$\exp(\theta_3 + \cos(\theta_4 x_3))$	34.6	1496	1494	719
$\theta_6 x_5$	28.9	1244	1243	609

subexpression can be removed without severely reducing the goodness-of-fit. The result is similar to Table 5.4 but here the analysis is not limited to nested models but includes larger subexpressions.

For fitting the coefficients of reduced models, the coefficients of the full model and the average evaluation result of the removed subexpression can be used to calculate the starting point for optimization of coefficients. The procedure produces an importance value for each of the subexpressions, whereby the leaf nodes will tend to have smaller importance, and importance increases when moving up to the root node.

The importance of subexpressions can be visualized using an expression tree as for example shown in Figure 5.10. This can guide manual simplification of expressions. This approach to model simplification is also called *pruning* in GP literature.

The method for the calculation of subexpression importance is computationally costly as it involves running an iterative algorithm for nonlinear optimization for each tree node. Additionally, the importance values have to be recalculated when a subtree is removed because importance values of subtrees are not independent. We therefore recommend to use this method mainly for the post-hoc simplification of expression trees produced by GP. Pruning of solution candidates could also be helpful within GP to open up potential extension points to include more relevant subexpressions from other solution candidates. Selective application of pruning e.g., only for the fitter or older solution candidates can be worthwhile to analyze in more detail. In any case the application of pruning within the genetic loop requires a more efficient way to quickly approximate the importance of subexpressions. Ideally, one can reuse information that is anyway available from the evaluation of solution candidates to estimate the importance.

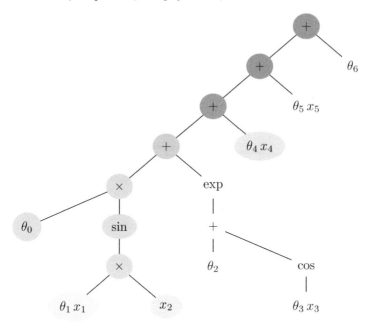

Figure 5.10: Example for the visualization of subexpression importance for an expression tree.

5.4 Example: Boston Housing

We have discussed several techniques for model validation, simplification, and selection and used a synthetic dataset to demonstrate the techniques. In this section we use them for an actual modelling task. The Boston housing dataset is a great example for this purpose. It's a medium sized dataset with originally 509 rows and 14 variables and exhibits several characteristics that are typical for real-world datasets. This dataset was published originally in a 1978 paper on "the willingness to pay for clean air" (Harrison and Rubinfeld, 1978) and became a well-known benchmark dataset through its inclusion in the StatLib repository.[2] Harrison and Rubinfeld (1978) developed a regression model mainly with the aim to check if there is a statistical dependency between the NO_X concentration – a surrogate for air pollution – and median housing values in Boston towns.

This dataset was collected almost fifty years ago in the context of a completely different society which is apparent by the fact that it includes a feature for the "black proportion by town". The Boston housing dataset is interesting

---

[2]http://lib.stat.cmu.edu/datasets

Table 5.6: Variables in the housing dataset.

CRIM	per capita crime rate by town
ZN	proportion of residential land zoned for lots over 25,000 sq.ft.
INDUS	proportion of non-retail business acres per town.
CHAS	Charles River dummy variable
	(1 if tract bounds river; 0 otherwise)
NOX	nitric oxides concentration (parts per 10 million)
RM	average number of rooms per dwelling
AGE	proportion of owner-occupied units built prior to 1940
DIS	weighted distances to five Boston employment centres
RAD	index of accessibility to radial highways
TAX	full-value property-tax rate per $10,000
PTRATIO	pupil-teacher ratio by town
B	$1000(\mathrm{Bk} - 0.63)^2$ where Bk is the proportion of black
	population by town
LSTAT	percentage of lower status of the population
MEDV	median value of owner-occupied homes in $1000's

for historic reasons because it has been used for such a long time and so widely for benchmarking purposes in ML.

Many researchers used this dataset since then to demonstrate the capabilities of different statistical modelling and ML methods. It became popular probably because it was one of the few datasets on housing prices of moderate size available on the early Internet (Gilley and Pace, 1996). In the cases where the dataset is used for benchmarking purposes, usually only the predictive accuracy of the model is measured. This must be distinguished from the original purpose which was statistical estimation of effects on house values and not prediction. Harrison and Rubinfeld (1978) manually defined features based on domain knowledge and focused on the interpretation of the estimated coefficients. When using SR for this example, we are also interested in interpretation of the model but we do not use domain knowledge beyond the features that are available in the published dataset. The 14 features collected in the Boston housing dataset are shown in Table 5.6.[3]

We use the dataset here to demonstrate model selection and validation techniques for SR. We generate a model for the median value of homes (MEDV) and compare our model to the Harrison-Rubinfeld model. As a second example we generate a model for the NO_X concentration.

5.4.1 Data Preprocessing

The original dataset published by Harrison and Rubinfeld (1978) has several incorrect values for MEDV which were later identified and corrected (variable:

---

[3]Taken from `https://www.cs.toronto.edu/~delve/data/boston/bostonDetail.html`

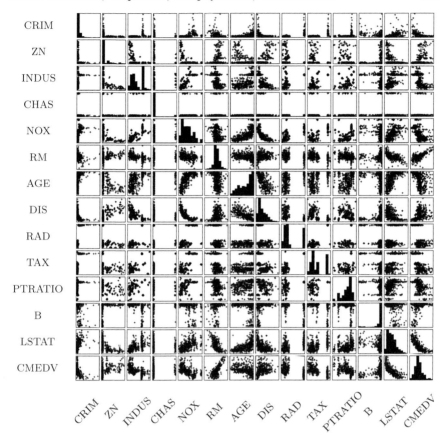

Figure 5.11: Scatter plot matrix for the corrected Boston housing dataset.

CMEDV) by Gilley and Pace (1996). We use the corrected dataset which includes the latitude and longitude of census tracts (Pace and Gilley, 1997) as well as the name of the towns for each census tract (observation). An additional issue is that for several observations the median value is exactly $50,000 which is at the same time the maximum value. This is an artifact from limiting the maximum value to $50,000 (Gilley and Pace, 1996). Since we do not know the correct value for these observations we removed those observations from the dataset.

Figure 5.11 shows the scatter plot matrix for the housing dataset with a small amount of jitter added to the values to show duplicate data points in the chart. The empirical distributions of most variables differ from a Gaussian distribution. LSTAT and RM are correlated with MEDV and with each other. AGE and DIS are correlated with NOX.

The data is structured hierarchically (Bivand, 2017). Each observation contains aggregate values for one census tract including the median value of

homes. Each census tract is part of a town whereby the number of tracts
within a town is variable. Several of the variables, in particular: ZN, INDUS,
NOX, DIS, RAD, TAX, and PTRATIO are only known on the level of towns
(approximately). The variables with census tract resolution are: MEDV, CRIM,
CHAS, RM, AGE, B, and LSTAT. The reason for the different resolution levels
is that the data were combined from different data sources. Bivand (2017)
discusses the impact of the different resolution levels in detail and shows that
copying out the features from the higher-level zones to the census tracts severely
affects modelling results. He recommends to instead aggregate the census tract
data to the zones to improve the model as this reduces spatial correlation
between observations and also improves heteroscedasticity. For our example
model for MEDV, we use the fine-grained data anyway, but carefully partition
the dataset into training and testing sets by grouping observations by towns.
Traditionally, a random assignment of observations is recommended, but this
would be problematic here because observations are not independent because
of the spatial correlation and random assignment to folds leads to undetected
overfitting. Therefore, we group observations by towns and randomly assign
these groups to partitions and cross-validation folds. This is comparable to
modelling based on aggregated data as recommended in Bivand (2017), but
alleviates the issues only partially as this way of grouping has the effect that
towns which are split into more census tracts have more weight in the model.

We assigned towns to training and testing partitions randomly to achieve
almost a 66/34% split. Figure 5.12 shows the locations of census tracts assigned
to the training and testing sets. The visualization shows that most tracts lie
in the center of Boston which is split into many census tracts. The grouping
of observations is clearly visible. The towns in the training set were similarly
partitioned into five subsets of almost equal size for cross-validation.

As a final preprocessing step we calculate log-transformed values for the
corrected MEDV values as in Harrison and Rubinfeld (1978). One argument
for using this transformation is that MEDV must be positive, and this is
guaranteed after back-transformation of the predictions. Another argument is
that the prediction model should have small relative errors instead of small
absolute errors.

5.4.2 Model Generation and Selection for Median Values of Homes

Table 5.7 shows the GP parameters whereby we vary the maximum tree size
and use 10-fold CV to find the best value. The CV results shown in Figure 5.13
imply that we should use a small maximum tree size. The best CV score is
found for a maximum size of seven nodes. The CV scores have large standard
errors and the configuration with only a maximum of three nodes[4] has a CV

---

[4]Operon counts a variable with its coefficient (coeff × var) as a single node.

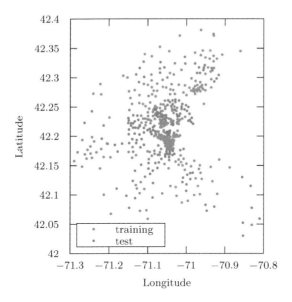

Figure 5.12: Geographical location of census tracts and their grouped assignment to training and testing sets by town.

score almost within one standard error. Based on the CV results we should select either three nodes or seven nodes as the maximum tree size.

To give an impression of how well the CV score correlates with the test error in this example, we also show the average test MSE of the ten models in the plot. Here, the test scores do not correlate well with the CV scores. The best average test RMSE is 0.1957 and is reached with maximum size 15. The average test RMSE for seven nodes is significantly higher (0.2264). Thus, the cross-validation approach does not work well in this example to select the best size limit.

To demonstrate the effectiveness of the model selection criteria for SR, we run GP with the settings shown in Table 5.7 ten times for the 11 size limits, collect all 110 models, and simplify and optimize the parameters (see Section 6.2). Then we calculate AICc, BIC, and DL for all models with $\sigma_{\text{err}} = 0.175$ which was estimated via random forest regression. Table 5.8 shows the number of parameters as well as the training and test RMSE for the best test model, as well as the models with best AICc, BIC, and DL values. Note that the models selected by all three criteria have similar test RMSE, whereby the model with smallest DL is best and has fewest parameters. The MDL model is shown in Figure 5.14 together with a visualization of the importance of subexpressions. The subexpression importance shows that all four terms are important. None of the nodes can be removed without increasing the training error significantly.

Figure 5.15 shows the scatter plot of predicted over target values and

Table 5.7: GP parameter settings for finding a prediction model for log(CMEDV).

Parameter	Value
Population size	1000
Generations	100
Max. size	$\{3, 5, 7, 10, 15, 20, 25, 30, 35, 45, 55\}$ nodes
Local opt.	10 iterations Levenberg-Marquardt
Fitness	negative MSE on training set (maximized)
Selection	tournament with group size 5
Mutation rate	25%
Function set	$+, \times, \div, \log(x), \sqrt{x}$
Terminal set	{ coefficient, coefficient $\times$ variable }
Input variables	CRIM, ZN, INDUS, CHAS, NOX, RM,
	AGE, DIS, RAD, TAX, PTRATIO, LSTAT
Target	log(CMEDV)

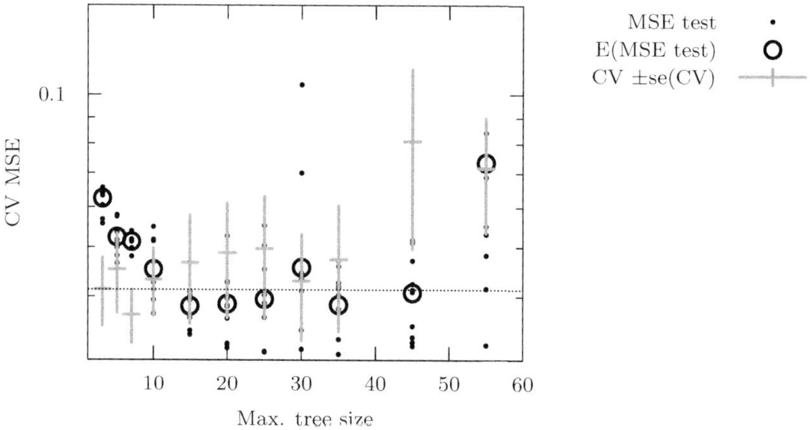

Figure 5.13: Cross-validation results for the prediction of log(CMEDV) over maximum tree size. Here CV does not work well to select the best model size. The best CV MSE is reached with seven nodes, while a limit of 15 nodes works best on average for the test set.

Table 5.8: Prediction errors of models for log(CMEDV) selected by AICc, BIC, and DL compared to the model with best RMSE on the test set.

Criterion	Num. Param.	RMSE (train)	RMSE (test)	Rank (test)
RMSE (test)	18	0.1274	0.1754	1
DL	6	0.1537	0.1915	22
BIC	12	0.1336	0.1939	25
AICc	17	0.1234	0.1961	32

$$\log(\text{CMEDV}) \approx \theta_1 \text{LSTAT RM} + \theta_2 \text{RM} + \theta_3 \text{RM}^{-1}$$
$$+ \theta_4 \frac{\text{LSTAT}}{\text{PTRATIO} + \theta_5 \text{CRIM}} + \theta_6 \qquad (5.23)$$
$$\theta = (-9.476 \times 10^{-3}, 0.587, 13.02, 0.649, 1.567, 2.410)$$

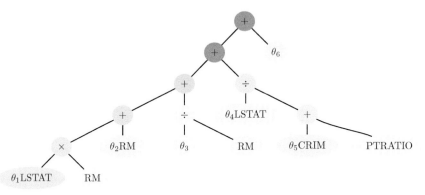

Figure 5.14: Subexpression importance for the simplified model for log(CMEDV).

clearly shows a heteroscedastic error distribution with larger variance for small CMEDV values. The observations with low CMEDV values affect the model strongly. We take note of this, but continue with the analysis of the selected model, because overall the scatter plots are similar for the training and testing partitions. Two options to handle this issue are to try other transformations of the target variable, or to use a different objective function.

The plot of residuals shown in Figure 5.16 allows us to check whether the model is biased, or failed to capture correlation between individual inputs and the target variable. As a visual aid to highlight deviations from zero, scatter plot smoothing results are also plotted. The only notable deviation from zero mean is for low TAX values, where the actual value is higher than predicted by the model. Overall however, the model has captured the uni-variate correlations reasonably well.

Finally, the intersection plot in Figure 5.17 shows the correlations captured by the model. The static plot is a bit limited because it does not show the interactions between LSTAT and RM and LSTAT, PTRATIO, and CRIM. Nevertheless, the plot allows us to interpret the correlations identified by the model. The model predicts a lower value for higher crime rate, whereby the effect quickly levels off for the few census tracts with extremely high crime rates. The predictions over the number of rooms (RM) is interesting, because the model predicts higher value for census tracts with low or high average number of rooms. One explanation for this could be that the four or five census tracts with low average room number occur primarily in the city center which

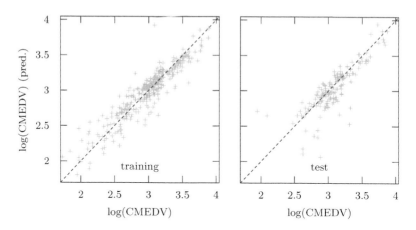

Figure 5.15: Scatter plots for the selected log(CMEDV) model (Equation (5.23)) on training and test sets.

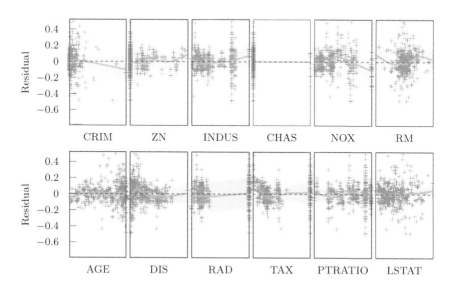

Figure 5.16: Residuals plot for the log(CMEDV) model.

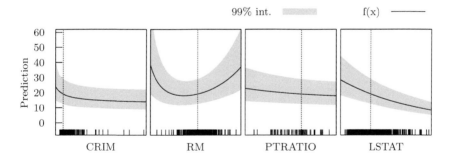

Figure 5.17: Intersection plot for the CMEDV model after back-transformation.

is more expensive. These census tracts are outliers that are also visible in the scatter plot of RM and CMEDV in the matrix shown in Figure 5.11. The model predicts a slightly lower value for tracts with higher pupil to teacher ratio and for tracts with a large ratio of lower status population. All of these correlations are plausible.

It is interesting to compare the model found with SR to other models described in the literature. Harrison and Rubinfeld (1978) used Equation (5.24) which uses all variables in the dataset.

$$\log(\text{MEDV}) \approx \beta_1 + \beta_2 \text{RM}^2 + \beta_3 \text{AGE} + \beta_4 \log(\text{DIS}) + \beta_5 \log(\text{RAD}) + \beta_6 \text{TAX}$$
$$+ \beta_7 \text{PTRATIO} + \beta_8 \text{B} + \beta_9 \log(\text{LSTAT}) + \beta_{10} \text{CRIM}$$
$$+ \beta_{11} \text{ZN} + \beta_{12} \text{INDUS} + \beta_{13} \text{CHAS} + \beta_{14} \text{NOX}^2$$
$$(5.24)$$

Fitting this model with ordinary least squares to the original dataset including censored data and incorrect values they achieved $R^2 = 0.81$ on the full dataset. Later Gilley and Pace (1996) corrected data for the target variable and removed the censored observations with MEDV = 50,000 and refitted the same model. They report $R^2 = 0.806$ for the original dataset and $R^2 = 0.811$ for the corrected dataset. The model was further extended to include the latitude and longitude of towns by Pace and Gilley (1997) with a marginal improvement to $R^2 = 0.814$.

The main objective of all of these models from literature was the estimation of the effects of the parameters in the dataset and not prediction. Our SR model is primarily a predictive model because we purposefully do not include all parameters and use a training/test split to allow estimation of the expected prediction error. Additionally, we selected the model using the MDL principle with the aim to select a model with good predictive performance.

Nevertheless, we can compare the SR model to the literature models for both scenarios, when we fit the model to the full dataset (for estimation), and when we use two separate partitions for fitting and evaluation to estimate the predictive performance. The results are shown in Table 5.9 which shows that

Table 5.9: Comparison of the SR model to the Harrison and Rubinfeld OLS model.

	Estimation		Prediction (train)		Prediction (test)	
Num. rows	490		328		162	
Model	OLS	SR	OLS	SR	OLS	SR
RMSE	0.168	0.165	0.159	0.154	0.204	0.191
R^2	0.807	0.814	0.850	0.860	0.529	0.600

the SR model has a better fit both for estimation and prediction even though it is much smaller than the OLS model.

Almost fifty years after the original paper, it is difficult to check whether the interactions captured by SR are meaningful. In any case, SR captured nonlinear dependencies and interactions between variables that were not considered in the original models. The SR model is smooth and predicts CMEDV in a hold-out set reasonably well. Interestingly, the concentration of NO_X – which was a main focus of Harrison and Rubinfeld (1978) – is not included in our SR model.

5.4.3   Model Generation and Selection for NO_X Concentrations

In this section we use SR to build a prediction model for NO_X, again applying the techniques introduced in this chapter. The NO_X values in the dataset stem from a simulation model (the Transportation and Air Shed SIMulation model (TASSIM)) calibrated to values from monitoring stations (Bivand, 2017). The NO_X values are calculated for larger geographical zones which match the towns approximately. We therefore use data aggregated on the town level for the NO_X model. Correspondingly, we have fewer observations in the dataset, and we use only the input variables ZN, INDUS, PTRATIO, RAD, TAX, DIS. These variables are on the same resolution at the level of towns, except for DIS, for which we use the mean value over the census tracts for each town. The assignment of towns to training and testing partitions and cross-validation folds is the same as for the MEDV model.

We use almost the same GP parameters as shown in Table 5.7, but stop already after 25 generations. The function set is $\{+, \times, \log(x), \sqrt{x}\}$. Again we run GP ten times for each configuration and collect the best model found in each run into a set of candidate models. Only 63 unique models for NO_X remain after simplification, parameter optimization, and removal of duplicates.

Again we use AICc, BIC, and DL to select three models from the set of unique models using the estimate $\sigma_{\text{noise}} = 5.25 \times 10^{-2}$. Table 5.10 shows the number of coefficients and RMSE values on training and testing partitions for the best models. The best model on the test set has 12 coefficients and RMSE=4.693×10^{-2}. This time the model selected with the AICc criterion

Table 5.10: Models for NO_X selected by AICc, BIC, and DL compared to the best model on the test set.

Criterion	Num. Param.	RMSE (train)	RMSE (test)	Rank (test)
RMSE (test)	12	3.423×10^{-2}	4.693×10^{-2}	1
AICc	9	3.361×10^{-2}	5.852×10^{-2}	31
BIC	3	4.472×10^{-2}	6.207×10^{-2}	37
DL	3	4.562×10^{-2}	6.989×10^{-2}	52

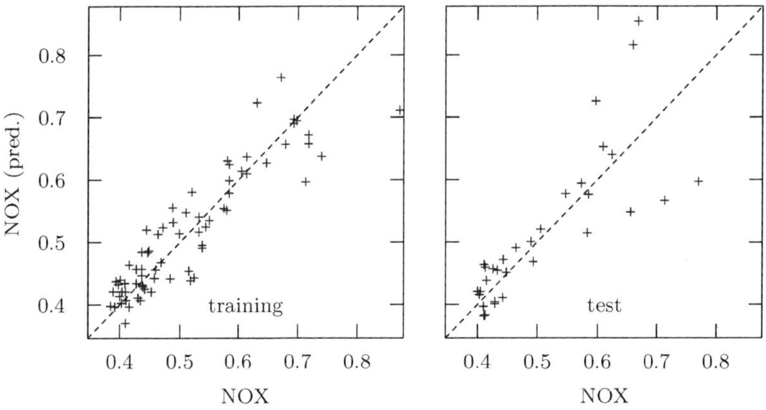

Figure 5.18: Scatter plots for the MDL model for NO_X (Equation (5.26)) on training and test sets.

with nine coefficients has the best test error (rank 31 out of 63). The best BIC model only has three coefficients and is only slightly worse (rank 37). The model with minimal DL also has only three coefficients but is the worst of the three models. The model with best BIC (Equation (5.25)) and the MDL model Equation (5.26) are similar; both use only a single variable, the distance to the five employment centers.

$$M_{\text{BIC}} : \quad \text{NOX} \approx -0.484 \log \text{DIS} + 0.314\sqrt{\text{DIS}} + 0.533 \qquad (5.25)$$

$$M_{\text{MDL}} : \quad \text{NOX} \approx -0.245 \log \left(\log \text{DIS} + 0.236 \right) + 0.627 \qquad (5.26)$$

The prediction errors of all models are relatively large on both the training and the testing sets. Figure 5.18 shows the scatter plot for M_{MDL} as an example. The distribution of residuals does not differ notably between training and testing partitions.

The intersection plot for the NO_X model in Figure 5.19 visualizes the dependency of DIS (weighted distance to five Boston employment centers) and NO_X as captured by the MDL model. It is plausible that air pollution is concentrated in areas of higher traffic. The model predicts that NO_X is higher

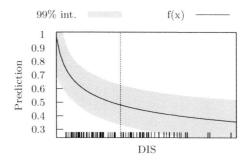

Figure 5.19: Intersection plot for the MDL model for NO$_X$.

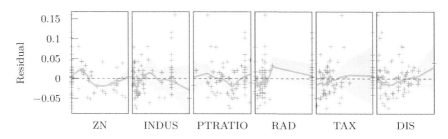

Figure 5.20: Residuals plot for the MDL model for NO$_X$.

close to employment centers and gets smaller with increasing distance. This correlation can be observed in the scatter plot matrix in Figure 5.11 as well.

The residuals plot for the MDL model in Figure 5.20 shows no clear patterns remaining in the residuals, which is a hint that including DIS in the model is sufficient.

In summary, for the housing example cross-validation was less effective than for the synthetic Friedman dataset. Even though we were careful in partitioning the data, the spatial correlation and the hierarchical nature of the data made it difficult to create independent training and testing partitions. As a consequence the correlation of the CV score with the test error was low. Additionally, heteroscedasticity and high noise introduced difficulties. Nevertheless, the model selection criteria AICc, BIC, and DL worked quite well and selected models that performed well on the testing sets. For the median value of homes (MEDV) we found a short interpretable and plausible nonlinear model that performed better than the linear models described in early literature and is much less complex.

5.5 Conclusions

As SR is essentially a search method for finding nonlinear regression models, we can use established techniques for validation of nonlinear regression methods. We recommend especially the visualizations of residuals over input variables, partial dependence plots, and intersection plots to understand, validate, and explain the nonlinear dependencies captured by the model. Often such plots together with the relative importance of each feature for the prediction are the most important output of SR projects. Visualizations such as partial dependence plots allow us to explain and understand how the model reponds to changes of inputs and increase the trust in the models identified by SR algorithms.

Depending on the configuration, GP may generate equations with deeply nested functions, redundant coefficients, and subexpressions that have almost no effect on the prediction. We therefore recommend to always simplify the generated equations ideally with a computer algebra system. Additionally, the importance of all subexpressions of the equations should be determined to find parts that can be removed with minor effect on the goodness-of-fit. Removing those less important subexpressions often simplifies SR models found by GP significantly.

Using a nondeterministic algorithm such as GP for SR allows to find several alternative models that fit the data well. This not only gives us a better feeling for the number of alternative models, but also allows to select one of the models that is preferable. Similarly, when using a multi-objective algorithm, we may select only a few models from the Pareto-front. The model selection techniques discussed in this section allow us quantitatively compare model based on combined measures of prediction error and complexity and to select those models confidently that are also likely to generalize well to new observations. We found that selection based on description length is conservative but works well in general and can be recommended for SR.

5.6 Further Reading

Classical references in nonlinear regression modelling are *Nonlinear Regression and Its Applications* by Bates and Watts (1988) and *Applied Regression Analysis* by Draper and Smith (1998). Both books discuss fitting of linear and nonlinear regression models but also statistical techniques for model validation and comparison.

Model selection techniques and theoretical background for model selection are discussed in *The Elements of Statistical Learning* by Hastie et al. (2001).

This book contains a good description of cross-validation, generalized cross-validation, and information theory-based model selection criteria.

A good book that gives background on information theory which underlies the model selection criteria discussed in this book is *Information Theory, Inference, and Learning Algorithms* by MacKay (2003). Akaike's information criterion and its relationship to the Bayesian information criterion is described at length by Burnham and Anderson (2003) in *Model Selection and Multimodel Inference: A Practical Information-Theoretic Approach*. The standard work for model selection by minimum description length is *The Minimum Description Length Principle* by Grünwald (2007).

The recent book *Interpretable Machine Learning – A Guide for Making Black-Box Models Interpretable* by Molnar (2020) discusses similar topics as this chapter and has more information on feature importance estimation. It also contains an understandable introduction to more recent techniques for explaining black-box models, namely, Shapley value additive explanations (SHAP) and local interpretable model-agnostic explanations (LIME).

6

Advanced Techniques

A variety of algorithmic extensions has been developed for SR over the last decades. Some of them have significantly increased the range of problems that can be solved, some of them made SR algorithms more efficient, more robust, or improved model accuracy. Some of those algorithmic extensions are described in this chapter.

Section 6.1 focuses on different techniques for integrating prior knowledge. In Section 6.2 we describe concepts for the optimization of model coefficients thus increasing model accuracy. The uncertainty of model predictions can be quantified using different approximations, which we describe in Section 6.3. SR can also be adapted to identify differential equations for describing dynamical systems (Section 6.4). In Section 6.5 we show how non-numeric data can be used with SR. Finally, we discuss non-evolutionary algorithms for SR in Section 6.6.

6.1 Integration of Knowledge

All supervised learning methods use data to learn the unknown dependency between independent variables and the dependent variable. Often there is additional knowledge or side information which helps to describe this dependency. In such cases it is important to consider how this information can be used in combination with the data to improve the model or the efficiency of the learning process. Side information can for instance be useful to reduce overfitting or to improve the extrapolation capabilities of models. The most common way to integrate prior knowledge is to define a problem-specific model structure. Especially in SR, the model representation in the form of mathematical expressions opens up intuitive approaches that allow to hybridize knowledge- and data-based approaches. For example, models that predict system behaviour well, even in areas where there are only limited data, or when extrapolating, can be generated with the help of *shape constraints* or model *structure templates*.

Knowledge integration is especially relevant when we use supervised learning methods in the field of natural sciences. Here many of the main mechanisms are often well understood and described by physics-based models which allow extrapolation. This is in contrast to purely empirical data-based models which are valid only within the domain of the observations used for training the model. However, data-based models are often useful as surrogates for physics-based models for phenomena that are difficult to describe exactly or computationally very expensive.

In the following sections, we discuss some ways how data-based and physics-based modelling approaches can be combined. We focus on SR as a supervised learning approach and discuss how knowledge e.g., in the form of physics-based models can be used in this process.

6.1.1 Example Applications

We find many examples in this book where we use a data-based approach for learning but at the same time have side information that can be used to improve the model. Consider for example the prediction of atmospheric CO_2 concentration in Section 7.8. In this example we have a long time series of observations that clearly shows a periodic pattern as a result of the solar cycle. Even if we would have only one or two years of observations, we would know that the time series should be periodic and could use this knowledge to ensure that all models exhibit this periodic pattern even when forecasting into the future. We will later discuss how this can be accomplished in the case of SR.

In Section 3.2 we gave the example of modelling dynamics of different types of balls dropped from a bridge, where we have observations only for a height which is not sufficient for the balls to reach terminal velocity. However, as we know that the velocity will reach a limit because of drag if the ball is dropped from a higher position, we could try to enforce that the SR model has this property. This would allow us to improve the extrapolation capabilities of the model.

Another example where integrating knowledge can be beneficial is the friction model in Section 7.4. The mechanics of the system are well-described by physical laws. The friction force however is a phenomenon that arises from microscopic forces between the sliding surfaces and there is no formula to calculate it exactly. Friction depends on many variables, including the surface properties, forces between the surfaces, and sliding velocity and can be described on a macro-level only empirically. By combining physical laws with the empirical friction model, we could simulate the mechanics of a larger system.

6.1.2 Knowledge Integration Methods

The inclusion or exclusion of functions and operators in the function set is a simple way of knowledge integration. If we know that a model must be periodic, we should include periodic base-functions e.g., trigonometric functions. If we know that the model must be bounded, we should allow fractions or bounded functions such as *tanh* because expressions generated using the limited operator set $\{+, -, \times\}$ are polynomials which are unbounded.

We distinguish knowledge integration methods into two categories: those which act through the fitness function, and those which act through the hypothesis space. The former category contains all methods which use customized fitness functions to integrate knowledge and therefore implicitly affect selection and replacement. All methods that affect the possible size or structure of expressions fall into the later category.

6.1.3 Knowledge Integration via Customized Fitness Evaluation

The fitness function is the most obvious point where side knowledge can be used in GP. In the fitness function for SR we primarily use data to calculate the prediction error of the model. Any information that we have additionally can be used in combination with the data to determine fitness. As a consequence fitness and therefore selection and replacement probability is not solely determined by the data alone and an individual with lower prediction error may be assigned a worse fitness value than one with a higher prediction error if it has other beneficial traits with respect to side knowledge. In other words, individuals with lower prediction error are not always preferred.

Multi-objective algorithms can be recommended to handle multiple contradictory fitness criteria. On the one hand, we want to optimize the fit to the data which means minimizing error. On the other hand, we want to ensure that the function conforms to our knowledge of the observed system, for example that the model is monotonic or symmetric.

In a single-objective algorithm we may combine multiple fitness criteria into a single value, or we may simply reject solution candidates that do not conform to existing knowledge. Another option are evolutionary algorithms that explicitly support constraints (see Coello Coello (2002) for a survey). These algorithms are preferred if we want to find only models that conform to constraints instead of searching multiple models that tradeoff multiple fitness criteria.

6.1.3.1 Preventing Singularities

SR may produce model functions that are only partially defined or have singularities, for example from division by zero or calculating the logarithm or square root for a non-positive argument. An example for a function with a singularity at $\frac{1}{3}$ is shown in Figure 6.1. Without specific measures to protect against this, we may not detect such issues if they do not occur for the observations available in the training phase. However, in many applications, we want to prevent singularities already in the training phase or when selecting models. This is not straightforward in general. Obviously, if the expression does not contain a division operator we can assume that there will be no division by

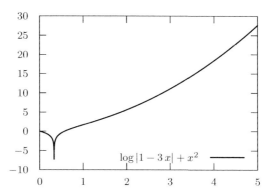

Figure 6.1: Example function with a singularity at $\frac{1}{3}$.

zero. In contrast, if we find the model contains a division operator, we have to analyze whether the denominator can become zero for any allowed input. This is NP-hard for nonconvex functions (Murty and Kabadi, 1987). One option is to use sampling to find singularities because we only need to check the model response and do not need to compare to a target value. But sampling is inefficient especially for models with multiple input variables because the number of samples required to cover the input space well grows exponentially with the dimensionality of the problem.

A simple solution to reliably detect all singularities within the input space has been proposed by Keijzer (2003). He introduced the idea to use interval arithmetic, usually used for determining numerical precision of floating point code (Alefeld and Mayer, 2000), to calculate outer bounds for the model response over the whole input space. If the bounds indicate that there is a potential singularity in the expression, we assign a low fitness to that individual to prevent that it is selected for generating new individuals.

The evaluation rules for interval arithmetic are shown in Table 6.1 and guarantee that the actual value is within the resulting interval. The calculation of the interval requires a single post-order iteration through the expression tree, whereby the intervals for inputs can be determined from the training data. To implement additional rules it is only required to find the range of possible values for the function depending on the argument. Monotonic functions such as $\exp(x)$ are very easy to handle. Some complexities arise for handling floating point semantics and rounding exactly. However, for an approximation method such as SR these details are less important.

The attractiveness of interval arithmetic is that it is very easy to implement, and computationally efficient because it requires only a single evaluation with intervals. However, it is not a perfect solution, as it ignores correlations between variables and subexpressions, and as a consequence may produce extremely wide bounds. It may also cause false positives for expressions such as $1/(x{\cdot}x{+}1)$ with $x \in [-1 \ldots 1]$ which does not have a singularity. This means that checking

Table 6.1: Interval arithmetic evaluation rules.

Symbol	Interval
variable	$[\min(\text{variable}), \max(\text{variable})]$
constant	$[\text{constant}, \text{constant}]$
$[a_l, a_h] + [b_l, b_h]$	$[a_l + b_l, a_h + b_h]$
$[a_l, a_h] - [b_l, b_h]$	$[a_l - b_h, a_h - b_l]$
$[a_l, a_h] \times [b_l, b_h]$	$[\min(a_l b_l, a_l b_h, a_h b_l, a_h b_h),$
	$\max(a_l b_l, a_l b_h, a_h b_l, a_h b_h)]$
$[a_l, a_h] \div [b_l, b_h], 0 \notin [b_l, b_h]$	$[a_l, a_h] \times [\frac{1}{b_l}, \frac{1}{b_h}]$
$[a_l, a_h] \div [b_l, 0]$	$[a_l, a_h] \times [-\infty, \frac{1}{b_l}]$
$[a_l, a_h] \div [0, b_h]$	$[a_l, a_h] \times [\frac{1}{b_h}, \infty]$
$[a_l, a_h] \div [b_l, b_h], 0 \in [b_l, b_h]$	$[-\infty, \infty]$
$\log([a_l, a_h]), a_l > 0$	$[\log(a_l), \log(a_h)]$
$\log([0, a_h])$	$[-\infty, \log(a_h)]$
$\log([a_l, a_h]), a_l < 0$	undefined
$\exp([a_l, a_h])$	$[\exp(a_l), \exp(a_h)]$

SR models with interval arithmetic may lead to rejection of good models and premature loss of diversity. Several techniques have been proposed for improving the accurary of intervals including affine arithmetic (de Figueiredo and Stolfi, 2004), monotonic extensions (Hansen, 1997), Taylor models (Berz and Hoffstätter, 1998; Berz and Makino, 1998; Berz et al., 2001), or modal arithmetic (Gardeñes et al., 2001) but those improved methods are hardly mentioned in SR literature, more difficult to implement, and computationally more demanding.

Taking the idea further, we can use the same approach to check if SR models are differentiable over the whole input space. This allows us to enforce smoothness of the function. Here we benefit from the fact that we work with symbolic expressions, because this allows us to build the derivatives symbolically. The derivatives can then again be checked using interval arithmetic. This leads to the idea of shape constraints which is discussed in the next section.

6.1.4 Shape Constraints

The problem of unwanted singularities highlights that the goodness-of-fit on a training sample is often not sufficient for model selection. For example, given two regression functions with similar prediction error we prefer the smoother (less complex) function. However, it is difficult to quantify the smoothness of a function over the whole input space. *Shape constraints* allow to restrict the shape of the regression function and thus to integrate domain knowledge (Wright and Wegman, 1980; Kronberger et al., 2022; Curmei and Hall, 2023). For example to express that a univariate model $f(x)$ has to be monotonically increasing over variable x in a given range we would accept only those $f(x)$

Table 6.2: A list of some properties of functions that can be expressed as shape constraints.

Property	Constraint
	$\forall_{\mathbf{x}\in\Omega}$:
Boundedness	low $\leq f(\mathbf{x}) \leq$ high
Monotonicity over x	$\frac{\partial f(\mathbf{x})}{\partial x} \geq 0$
Convexity over x	$\frac{\partial^2 f(\mathbf{x})}{\partial x^2} \geq 0$
Smoothness over x	$\lVert \frac{\partial f(\mathbf{x})}{\partial x} \rVert_2 < s$
Symmetry of x_1, x_2	$f(x_1, x_2) - f(x_2, x_1) = 0$
Even in x	$f(x) - f(-x) = 0$
Odd in x	$f(x) + f(-x) = 0$

for which

$$\frac{\mathrm{d}f(x)}{\mathrm{d}x} > 0, x \in [x_{\text{low}}, x_{\text{high}}]$$

The main difficulty when evaluating shape constraints for the infinite input range $x \in [x_{\text{low}}, x_{\text{high}}]$ is the calculation of lower and upper bounds for $f(x)$ for all x in the input space. Interval arithmetic allows to approximate these bounds efficiently.

Shape constraints allow to enforce other properties additionally to monotonicity which are shown in Table 6.2. Usually, we combine multiple shape constraints to express the prior knowledge about the function to be identified. Additionally, we may use different constraints on different subsets of the input space Ω.

Shape constraints can help to stabilize the model in settings with high noise levels or to improve extrapolation behaviour. The following two examples help to demonstrate the effect. First, let us consider the example shown in Figure 6.2. It shows noisy data sampled from the function x^3. Since there are only a few points in this dataset and we have a low signal-to-noise ratio, we must control model complexity and fit a simple model. Correspondingly, we set the maximum length of the model to ten nodes, and use the function set $\mathscr{F} = \{+, \times\}$ with coefficients and variables in the terminal set. Correspondingly, the SR model will be a polynomial with small degree. Figure 6.2a shows the predictions of models found in 30 SR runs. In each run we randomly selected a subset of two-thirds of the observations for training to show the variance over training samples. Because of the noisy observations most of the models are not monotonic. Now assume that we have side knowledge that the data stem from a monotone process, and we therefore use the constraint $\frac{\partial f(x,\boldsymbol{\theta})}{\partial x} > 0$ to enforce monotonicity. This has the effect that models are closer to the generating function as shown in Figure 6.2b, and therefore also have lower expected prediction error.

Now consider another example shown in Figure 6.3 where training data are sampled from a sigmoid function again with additive Gaussian noise. Note

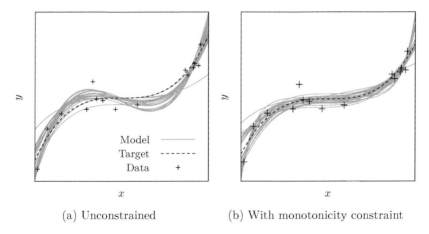

(a) Unconstrained (b) With monotonicity constraint

Figure 6.2: Demonstration of shape constraints to improve model quality for highly noisy data.

that the points are sampled only from a window in the center of the plot. With SR we can find models that fit the training data well. Extrapolation however is not feasible as shown in Figure 6.3a. If we know that the generating process is bounded from above and below, we can use three constraints: $f(x, \boldsymbol{\theta}) > 0$, $f(x, \boldsymbol{\theta}) < 1$ and $\frac{\partial f(x, \boldsymbol{\theta})}{\partial x} > 0$, $x \in [-10..10]$. The shape-constrained SR models shown in Figure 6.3b fit almost perfectly to the generating function even when extrapolating. Of course, this requires that the function set includes functions that are bounded such as a logistic function or *tanh*.

Figure 6.4 shows a zoom of Figure 6.3 focusing on the origin. Several of the models found without constraints are non-monotonic because of the noisy observations (Figure 6.4a) while enforcing constraints ensures that all models are monotonic (Figure 6.4b).

6.1.5 Knowledge Integration via the Hypothesis Space

SR methods usually have a parameter to control the size of the expression as well as the function set because these are the two most important configuration options that allow to adjust the algorithm for a particular problem. Setting the size limits and choosing the elements for the function set are so natural in the application of SR that one would not really think about this as knowledge integration. Instead we focus on methods where we help the algorithm by providing larger building blocks for the model. Here GP and especially grammar-guided forms of GP, for instance grammatical evolution, are especially useful because they allow to provide predefined expression fragments or limit the expression structure easily.

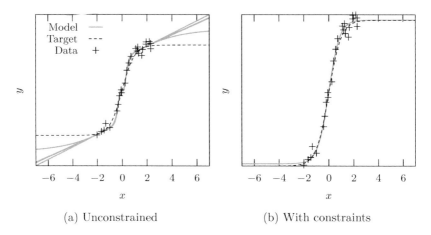

(a) Unconstrained   (b) With constraints

Figure 6.3: Demonstration of SR with shape constraints for improving extrapolation.

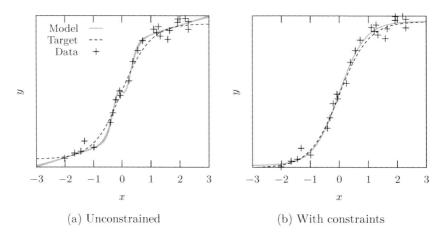

(a) Unconstrained   (b) With constraints

Figure 6.4: Demonstration of SR with shape constraints for enforcing monotonicity.

6.1.5.1 Problem-specific Functions and Terminals

Closely related to the configuration of the function set is the extension of the function and terminal sets with problem-specific building blocks that are thought or known to be necessary for a well-fitting model. Consider for example the periodic time series shown in Figure 6.5. It shows the water level at a measurement station in Elfin Cove, Alaska operated by the National Oceanic and Atmospheric Administration (NOAA) over three months. Since we know that tides are affected by gravitational forces from the position of the moon and the sun relative to the measurement station, we know that there must be at least three periodic components with known frequency. The first frequency is daily variation from the rotation of earth, the second frequency is the lunar cycle, and the third frequency is the solar cycle. Correspondingly, we may add three building blocks $\cos(f_1 x), \cos(f_2 x), \cos(f_3 x)$ to the terminal set to make it easier to find a well-fitting model.

In this example we can identify these frequencies from Figure 6.5 without knowing anything about the underlying physics. Alternatively, we may transform the time series to the frequency domain and estimate the most important harmonics from the spectrum. It is however not always so easy, for instance when additional variables have to be considered, when the time series is non-stationary, or when observations are less precise.

Using the harmonics as additional terminals for SR, can be accomplished by simply extending the dataset with generated variables. To demonstrate the idea we take the water levels reported every hour from May 2022, and train a model to predict hourly water levels for June 2022. For this short time period, the rotation of Earth around the sun has almost no effect. Therefore, we only consider the harmonics for the solar and lunar cycle, and add the calculated features $S_2 = t/24$, $S_1 = S_2/2$, $M_2 = t/24.8412$, $M_1 = M_2/2$, $\sin(2\pi S_2), \cos(2\pi S_2), \sin(2\pi S_1), \cos(2\pi S_1), \sin(2\pi M_2), \cos(2\pi M_2), \sin(2\pi M_1), \cos(2\pi M_1)$ to the dataset, where t is the number of hours since the start of the year. As function set we use $\mathscr{F} = \{+, \times, \div, \sin, \cos, x^2\}$. Other relevant parameters are the population size of 1000 individuals , fifty generations, a size limit of maximum fifty tree nodes, and local optimization of coefficients using a maximum of ten Levenberg-Marquardt iterations.

Figure 6.6 shows the line charts of the test MAE of the training best model in the population over the fifty generations from 30 independent GP runs with and without the additional features for harmonics. When using the additional building blocks it becomes much easier for GP to find a well-fitting model, because it does not have to identify these coefficients via fitting. The MAE on the test set is on average lower for the GP runs with the additional building blocks and the results are more reliable as the variance of test error over the runs is lower. Figure 6.7 shows for example the residuals of an SR model that uses the additional variables for the harmonics.

In conclusion, when you find yourself in a similar situation where the modeled process is known to be periodic, we recommend to add terminals

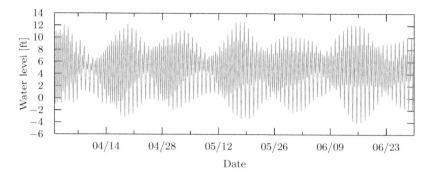

Figure 6.5: Water level measurements at Elfin Cove, AK for the Q2/2022 (source: NOAA www.noaa.gov).

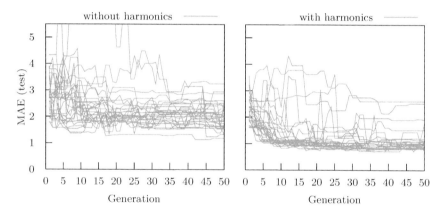

Figure 6.6: Line charts of test MAE over generations for 30 GP runs with and without harmonics in the terminal set. When using the predefined harmonics, GP finds better models more reliably (lower mean and variance).

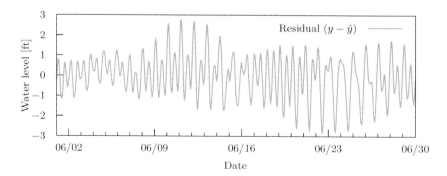

Figure 6.7: Residuals of the SR model using predefined harmonics for June 2022.

or functions with the correct frequency to improve modelling efficiency and reliability.

Another example where problem-specific physics-based building blocks are helpful is discussed in Section 7.10.4. Since we know for the pendulum and the double pendulum that the gravitational force is acting downwards we know that trigonometric functions are necessary to calculate the downward force from the angle. Correspondingly, we should include trigonometric functions in the function set to find a well-fitting model. For the double pendulum, we can also assume that the sum of the two angles must play a role, and include the building blocks $\sin(\theta_1 - \theta_2)$ and $\cos(\theta_1 + \theta_2)$. With this we increase the chance to identify approximations to the equations of motion from observational data and limited physical insight.

6.1.5.2 Structure Templates

In some applications, a system can be partially modeled using physical laws, while other effects have to be approximated either because it is computationally prohibitive to simulate or the underlying physics are unknown. When confronted with this situation, model structure templates can be used to combine the physics-based model with purely empirical models. Tree-based GP and especially grammatical evolution naturally allow to specify fixed parts of the model while leaving other parts open for evolution. Asadzadeh et al. (2021) use this approach for a hybrid (physics-based and empirical) model for a metal sheet-bending process. In the process it is useful to predict the stamping force

$$F = \frac{E}{E+h}\left(\frac{\sigma_y t^2}{w} + \frac{ht^3}{3Rw} - \frac{4\sigma_y^3 R^2}{3wE^2}\right)C \tag{6.1}$$

which depends on the sheet thickness t, the bending radius R, and material-dependent parameters describing the elastic-plastic deformation properties of the material, namely Young's modulus E, the yield strength σ_y, and the linear hardening modulus h. The three terms in Equation (6.1) represent the contribution of yield strength, plastic hardening, and elasticity, whereby the first term usually dominates the second term and the third term is several orders of magnitude smaller than the first term (Asadzadeh et al., 2021). Consequently, in many engineering applications the contributions of the second and the third term are ignored. This motivates the development of a hybrid model that can be fit to measurements whereby the first term is hardcoded in the model and the contribution of the other effects is only approximated via SR.

The hybrid model constains the known effect of yield strength as a fixed part, and has two placeholders for the approximation of elastic and plastic effects that shall be identified using GP-based SR as shown in Equation (6.2). The first placeholder f_1 represents the combined contribution of plastic hardening and elasticity. The second placeholder C represents a correction factor. For both placeholders, a function represented as a short expression is evolved and the whole model is fit to measurements.

$$F = \left[\frac{E}{E+h} \frac{\sigma_y t^2}{w} + f_1 \left(h, \sigma_t, t, \frac{R}{t}, \frac{w}{t} \right) \right] C \left(h, \sigma_t, t, \frac{R}{t}, \frac{w}{t} \right) \qquad (6.2)$$

As potential inputs to both functions the parameters h, σ_y, t, $\frac{R}{t}$, and $\frac{w}{t}$ are allowed. Using the dimensionless parameters $\frac{R}{t}$ and $\frac{w}{t}$ is again a form of knowledge integration via predefined building blocks and to improve scaling capabilities of the identified models.

There are multiple ways how similar hybrid models can be implemented depending on the capabilities of the GP software system that is used. One option is the use of tree-based GP with structure templates where the tree creation operator uses the predefined structure for all trees, and crossover and mutation operators are restricted to the subexpressions that are used for the placeholders. Another option is to use a grammar-guided GP system and fix the common structure in the grammar, allowing alternative subexpressions only for the placeholders. Yet another option that is available in most GP systems, is to implement a custom fitness function with the fixed part of the model hardcoded.

Asadzadeh et al. (2021) used tree-based GP with structure templates and demonstrated using a synthetic dataset that it is possible to find well-fitting and interpretable hybrid models. These models are short expressions using operators and functions that are available in all programming languages and can be easily implemented on programmable logic controllers (PLCs) for automated process control.

6.1.5.3   Dimensional Analysis and Dimensionless Parameters

Dimensionless parameters result from combining multiple parameters such that the units cancel out and can improve scalability of data-based models which enables to apply the same model to systems of different scale. In other words, using dimensionless parameters can improve extrapolation performance of fitted models. While SR is flexible enough to discover such dimensionless parameters automatically, it is more efficient to define such parameters manually if we know that they are relevant. Using dimensionless parameters is as simple as extending the dataset with the calculated parameters and adding them to the terminal set.

In the previous section we have already discussed an application where dimensionless parameters were used. Another example can be found in (Roland et al., 2019) where dimensional analysis in combination with SR is used to generate models for non-Newtonian flow in single-screw extruders.

6.2 Optimization of Coefficients

The abilities to perform implicit feature selection and to simultaneously evolve the structure and coefficients of the model represent one of the selling points for using GP for SR. Still, there are several requirements that have to be fulfilled by the final model so that it fits the data optimally:

1. The model uses an optimal subset of input variables (feature selection).

2. The model has an optimal structure for the selected subset of variables.

3. The model has optimal coefficients for the structure.

Only a model that fulfilles all three requirements can be detected as optimal.

Due to its stochastic nature it is common in GP that these three requirements are not fulfilled to the same degree. For example, a model may have an optimal structure, but does not fit the training data well due to inadequate coefficients. Linear scaling (Section 6.2.1) as well as memetic optimization techniques (Section 6.2.2) can be used to find coefficients more efficiently than by evolutionary means.

6.2.1 Linear Scaling

GP-based SR performs a parallel search of the space of models. Within this process, the fitness of solution candidates is evaluated using prediction error measures such as the sum of squared errors (SSE – see Equation (2.1)). However, GP requires a diverse and heterogeneous population with models that have responses on completely different scales than the target variable. With most of the common error metrics such models are assigned a low fitness. Keijzer (2004) motivates the need for scale-invariant fitness measures using the example shown in Figure 6.8. In this example we assume that the target variable y of a univariate regression problem is calculated as $y = 0.3\,x\,\sin(6.5\,x)$. Two models M_1 and M_2 are compared. M_1 is a constant model always predicting 0.0, while M_2 calculates its output as $\hat{y} = 1.2 + x\,\sin(6.2\,x)$. If the outputs of both models are taken as is (Figure 6.8a), the constant model M_1 has a smaller SSE than M_2, because it predicts the mean of y. Nonetheless, M_1 fails to capture the characteristics of y. If these two models were part of the GP population, the fitness and therefore the probability of M_1 to be selected for crossover would be higher. If the output values of both models are scaled to match the mean and variance of the target variable (Figure 6.8b), it becomes obvious that M_2 explains the variance in the data much better and only needs to be scaled correctly to have a low SSE. This is the main argument for using scale-invariant error measure for fitness evaluation.

This example illustrates an issue that is especially relevant for GP as it deals with two problems at the same time: a model with a structure that fits

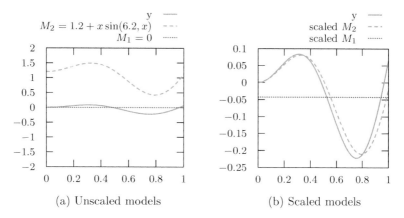

(a) Unscaled models   (b) Scaled models

Figure 6.8: Effects of linear scaling of two models M_1 and M_2 for the prediction of y. While the constant model M_1 has a lower error in the unscaled case (a), M_2 explains more of the variance in y when both models are linearly scaled (b).

the model – containing the relevant variables and their interactions – might get lost and removed from the population because of inappropriate coefficients, whereas a model with inferior structure might be preferred by the algorithm because it is closer to the mean of the target values.

For this reason, Keijzer (2003) introduced *linear scaling*. Linear scaling is defined as a linear transformation of the model output to match the variance and mean of the target variable. By using the squared Pearson's correlation coefficient (ρ^2) as a scale-invariant error measure for fitness evaluation, models that have a higher correlation with the target variable are assigned a higher fitness value. This is equivalent to minimizing the SSE for linearly scaled models (Keijzer, 2004). The best models found with such a fitness function can be scaled explicitly by simply adding coefficients for the intercept and scale, α and β. The scaling coefficients are calculated from the model response $\hat{y}$ for training data and the corresponding target values y. In tree-based GP explicit scaling is accomplished with additional nodes for multiplication of scale and addition of the intercept added on top of the root node of the expression tree.

$$f(\boldsymbol{x}, \boldsymbol{\theta})_{\text{scaled}} = \alpha + \beta \, f(\boldsymbol{x}, \boldsymbol{\theta}) \quad \beta = \frac{\text{cov}(f(\boldsymbol{x}, \boldsymbol{\theta}), y)}{\text{var}(f(\boldsymbol{x}, \boldsymbol{\theta}))}, \quad \alpha = \bar{y} - \beta \, \bar{f}(\boldsymbol{x}, \boldsymbol{\theta}) \ (6.3)$$

Linear scaling improves GP-based SR significantly, as it removes the necessity to evolve the optimal scale and offset for the models while only adding a minimal overhead for calculating the scaling factors. Linear scaling is easy to use as it just requires to replace the SSE with ρ^2 in fitness evaluation. However, simple linear scaling can only shift and scale the model response but cannot correct the effects of nonlinear coefficients occuring in SR models. This is indicated in the example in Figure 6.8, where the scaled model M_2 does not achieve a perfect fit, because the coefficient inside the sine function. For the optimization of such coefficients we need more powerful iterative nonlinear optimization methods as detailed in the following section.

6.2.2 Nonlinear Optimization of Coefficients

So far, we have demonstrated the use of linear scaling to address situations where the range of the model response is different from the target range. Generally, it is even better to optimize not only the scale and offset coefficients, but *all* coefficients occuring in the model.

In GP, coefficients may be evolved via mutation which randomly increases or decreases coefficients. Additionally, crossover may assemble short subexpressions consisting only of coefficients to calculate required values from multiple coefficients. This means the evolutionary algorithm must simultaneously evolve the structure of the model together with its coefficients which is arguably more difficult than solely evolving the model structure. Overall, this can hinder the performance of GP.

In this context, hybridizing GP with local search methods for fitting the coefficients of models can help to retain promising model structures thus improving the success rate (Topchy and Punch, 2001; Kommenda et al., 2020). Such algorithms that combine evolutionary with local search are known as *memetic* algorithms (Moscato, 1989; Neri et al., 2011). Although the original memetic algorithm used hill climbing for local search, many other algorithms such as gradient descent, simplex, or even other metaheuristics such as evolution strategies or particle swarm optimization can be used.

When all the functions that can occur in the model structure are differentiable, local optimization can be efficiently solved using a gradient-based algorithm such as the Levenberg-Marquardt algorithm for nonlinear least squares, or LM-BFGS (Liu and Nocedal, 1989) for general smooth loss functions. Gradient-free optimization methods such as CMA-ES (covariance matrix adaptation evolution strategies (Hansen et al., 2003)) which perform a broader search for a global optimum instead of converging to the local optimum may also be used.

In the following we briefly summarize the principle of gradient-based local optimization algorithms for nonlinear least squares (NLS) using the nomenclature from Moré et al. (1980) and describe how it can be integrated into GP. The NLS optimization problem is

$$\arg\min_{\boldsymbol{\theta}} \|F(\boldsymbol{\theta})\|_2 \tag{6.4}$$

In the case of nonlinear regression for target $\boldsymbol{y}$ and inputs $\boldsymbol{X}$ we have $F(\boldsymbol{\theta}) = \boldsymbol{y} - f(\boldsymbol{X}, \boldsymbol{\theta})$. Local NLS algorithms approach this problem by starting with an initial point $\boldsymbol{\theta}_0$ and making steps $\boldsymbol{p}$ to find new points $\boldsymbol{\theta}_{i+1} = \boldsymbol{\theta}_i + \boldsymbol{p}$ with $\|F(\boldsymbol{\theta}_{i+1})\|_2 < \|F(\boldsymbol{\theta}_i)\|_2$. For this, a linear approximation can be used based on the assumption that $F(\boldsymbol{\theta} + \boldsymbol{p}) \approx F(\boldsymbol{\theta}) + J(\boldsymbol{\theta})\boldsymbol{p}$, where $J(\boldsymbol{\theta})$ is the Jacobian of $F(\boldsymbol{\theta})$. In principle this means that for each step we have to solve the linear least squares problem to find the step $\boldsymbol{p}$

$$\min_{\boldsymbol{p}} \|J(\boldsymbol{\theta})\,\boldsymbol{p} + F(\boldsymbol{\theta})\|_2 \qquad (6.5)$$

Trust-region algorithms such as the Levenberg-Marquardt algorithm improve the convergence rate by constraining the step-size

$$\min_{\boldsymbol{p}} \|J(\boldsymbol{\theta})\,\boldsymbol{p} + F(\boldsymbol{\theta})\|_2 \quad \text{s.t.} \ \|\boldsymbol{D}\,\boldsymbol{p}\|_2 < \Delta \qquad (6.6)$$

whereby the maximum step size Δ is automatically adapted by the algorithm and $\boldsymbol{D}$ is a diagonal scaling matrix.

Equation (6.6) is solved repeatedly to find the steps p until the algorithm converges to a local optimum.

To use iterative NLS within GP we initialize coefficients of the expressions in the initial population randomly and call the NLS algorithm before calculating the SSE for fitness evaluation. For local optimization we first extract the coefficients of the expression as a starting point for optimization, then execute a few iterations of NLS, and finally update the coefficients if SSE was improved by the procedure.

The ability to efficiently compute partial derivatives is important for the performance of gradient-based optimization methods. We distinguish between three main approaches for calculating derivatives:

- *Symbolic differentiation.* Derivatives are computed using algebraic rules. This approach usually does not scale well with expression complexity and is tedious and error-prone to implement. In terms of performance, this is usually the slowest approach.

- *Numerical approximation.* The derivative is approximated using finite differences. This method may be inaccurate and suffer from numerical stability issues. Additionally it does not scale well with the number of parameters.

- *Automatic differentiation.* This method is based on the chain rule for calculating derivatives (Griewank and Walther, 2008) and is an important enabling factor for fitting deep neural networks. Therefore, well implemented software libraries for automatic differentiation are nowadays readily available and can be utilized to efficiently calculate partial derivatives. The computational effort for reverse-mode automatic differentiation to calculate the partial derivatives for all fitting parameters is on average only slightly larger than an evaluation of the fitness function. The backpropagation algorithm used for training neural networks is a variant of reverse-mode autodiff.

Since we do not have to rely solely on crossover and mutation to find the appropriate coefficients, we can avoid situations where low fitness caused by

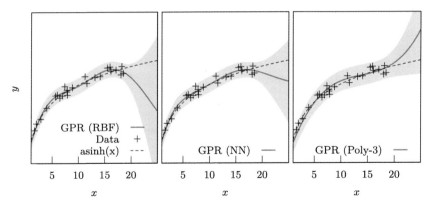

Figure 6.9: Example for three Gaussian process models for the *asinh* dataset.

ill-fitting coefficients prevents models with high adaptive potential from being taken into consideration by the selection mechanism.

In summary, gradient-based local search algorithms can use efficient procedures for computing the Jacobian using automatic differentiation. Since local search typically requires a number of iterations to converge, this increases the number of function evaluations within GP. However, despite this overhead memetic approaches have been shown to significantly improve algorithm performance (Z-Flores et al., 2014; Trujillo et al., 2018; Kommenda et al., 2020). Most modern SR algorithms use gradient information for coefficient adaption (La Cava et al., 2021).

6.3 Prediction Intervals

Prediction intervals are ranges within which the predicted value is likely to be located and allow to quantify the uncertainty of predictions. Wide prediction intervals indicate a high uncertainty of the model. A prediction interval combines the uncertainty which results from fitting models to limited data and uncertainty from the unexplained variation (noise) in the target values.

Prediction intervals are helpful because it is easier to make decisions when the certainty of the predictions is known. A model which is transparent and produces reliable prediction intervals, is more trustable and therefore preferable to models which produce only scalar predictions. Of course, this requires that prediction intervals correctly quantify the true interval of high probability for the predicted variable. Models with tighter prediction intervals and models with smaller prediction error are preferable.

Gaussian process regression (Rasmussen and Williams, 2006) is a well-

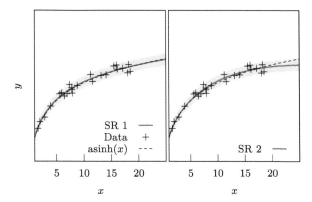

Figure 6.10: Two SR models with prediction intervals for the *asinh* example.

known Bayesian method that allows to calculate prediction intervals. It uses the kernel trick to fit a non-parametric nonlinear regression model. The choice of the kernel, or the covariance function, mainly determines the shape of the regression function and how well it can be fit to a given dataset. Each covariance function may have hyper-parameters which affect the shape of the regression function and have to be optimized for a given dataset.

To demonstrate the calculation of prediction intervals we use a synthetic dataset consisting of 30 samples (y, x) with $y = \mathrm{asinh}(x) + \varepsilon$, $x \sim U(0, 20)$, $\varepsilon \sim N(0, 0.05)$. Figure 6.9 shows three Gaussian process models for this dataset using three different covariance functions: radial basis functions (RBF), neural network (NN), and third-degree polynomial (Poly-3) (Rasmussen and Williams, 2006). The plots show that all three models match the data well. In the area of the input space that is well-covered by training samples the prediction mean is close to the generating function and the 95% prediction intervals accurately capture the noise in the data. When extrapolating for larger x values, however, the mean prediction has a larger error and the prediction intervals widen. The target function is contained in the prediction interval over the whole range used for x.

We expect prediction intervals to become wider when we make predictions far away from training data for example when extrapolating. The reason being that our uncertainty grows when we make predictions far from training data. For regions of the input space which are well-covered with training observations, the prediction intervals should be narrower. For a 95% prediction interval we expect that close to 95% of the observed values lie within the prediction intervals. This can be checked if the dataset is sufficiently large.

Prediction intervals for SR models can be calculated using well-known techniques for nonlinear regression models that we describe in more detail below. Figure 6.10 shows the predictions of two SR models with prediction intervals found for the *asinh* dataset. Compared to the Gaussian process models, we see that SR provides the potential to find models with improved

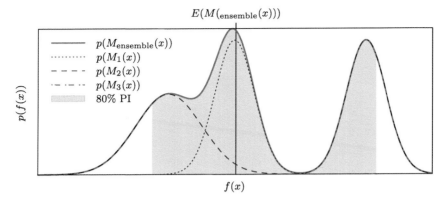

Figure 6.11: The combined predictive distribution of an ensemble of three models (M_1, M_2, M_3) may be multimodal with disconnected regions of high density. Here the mean prediction $E(M_{ensemble}(x))$ and the corresponding interval may not be representative for the models in the ensemble.

extrapolation capabilities and tighter prediction intervals. Here the reason is that SR has the chance to find a parametric expression that can be fit to the data compared to the three generic covariance functions used for the Gaussian process models.

There is another important conceptual difference in the prediction intervals for the Gaussian process models and the SR models. While the plots for the Gaussian process models show the posterior distribution of *all* functions that can be produced from the covariance function conditioned on the dataset, Figure 6.10 shows the predictive distribution only for two parameterized SR models found by GP. The prediction intervals for the SR models are calculated only from the uncertainty of the parameter estimates and the estimated noise variance. The structure of the model is assumed to be fixed and predefined. Alternative models that may also fit the data well but have not been selected by GP are ignored in these plots. It would also be possible to combine multiple well-fitting models with different structures found in multiple GP runs, or using multiple parameter vectors representing different local optima, into an ensemble. However, this most likely results in a multi-modal predictive distribution which is more difficult to work with. For example in a multi-modal density the mean prediction may lie between two modes and the corresponding prediction interval may include subintervals with small density as demonstrated in Figure 6.11. We will not go into further details about handling uncertainty in ensembles of SR models but instead focus on the estimation of prediction intervals for individual SR models.

Given a parametric model, we use the training set to determine which parameter values are most likely to produce the observations (maximizing the likelihood). The maximum likelihood estimate is a point estimate and

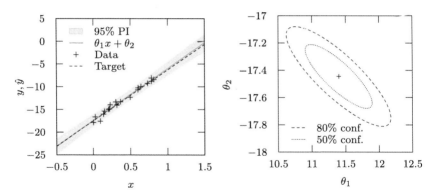

Figure 6.12: Example for prediction intervals and confidence regions (50% and 80%) for parameters for a linear model.

can be used to produce point predictions. For prediction intervals we have to determine regions in the parameter space of high likelihood. This can be difficult when the model function is nonlinear in the parameters. We describe ways to approximate confidence regions and prediction intervals which are often sufficiently accurate in the following.

6.3.1   Prediction intervals for Linear Models

Intervals for the expected response of linear models $Y = X\beta + Z$ are easy to determine exactly when we assume that Z is normally distributed and observed values of X are known exactly. The linearity implies a normal distribution for the predicted values Y as well as the estimated parameters β of the model. Correspondingly, putting a threshold on the likelihood imposes an ellipsoid contour in the parameter space containing all β values for which the likelihood is larger than the threshold. For the example shown in Figure 6.12 with two parameters θ_1, θ_2 the data and the model structure imply a two-dimensional Gaussian density for which the confidence regions for 50% and 80% are shown as elliptical contours in the right plot. The estimated distribution for the parameters together with the estimated noise variance then result in a predictive distribution which is again Gaussian for each input x as shown in the left plot. We first describe the calculation of intervals for linear models and then extend this for nonlinear SR models. The description follows (Bates and Watts, 1988).

When using the linear model $Y = X\beta + Z$ with $Z \sim_{iid} N(0, \sigma_{\text{err}})$, n observations the likelihood function is

$$\mathscr{L}(\boldsymbol{\beta}, \sigma_{\text{err}}) = \frac{1}{(\sigma_{\text{err}}\sqrt{2\pi})^n} \exp\left(\frac{-\|\boldsymbol{y} - \boldsymbol{X}\boldsymbol{\beta}\|_2^2}{2\sigma_{\text{err}}^2}\right) \qquad (6.7)$$

Assuming σ_{err} is fixed, the likelihood is maximized when the residual sum of squares $\|\boldsymbol{y} - \boldsymbol{X}\boldsymbol{\beta}\|_2^2$ is minimal. This optimization problem can be solved analytically and the maximum likelihood estimate $\hat{\boldsymbol{\beta}}$ is (assuming $\boldsymbol{X}^\top \boldsymbol{X}$ is invertible)

$$\hat{\boldsymbol{\beta}} = (\boldsymbol{X}^\top \boldsymbol{X})^{-1} \boldsymbol{X}^\top \boldsymbol{y} \tag{6.8}$$

A $1 - \alpha$ confidence interval for the expected response at a point $\boldsymbol{x}$ with k elements is

$$\boldsymbol{x}^\top \hat{\boldsymbol{\beta}} \pm s \sqrt{\boldsymbol{x}^\top (\boldsymbol{X}^\top \boldsymbol{X})^{-1} \boldsymbol{x}} \, t(n - k, 1 - \alpha/2) \tag{6.9}$$

where $t(n - k, 1 - \alpha/2)$ is the $1 - \alpha/2$ quantile for Student's t distribution with $n - k$ degrees of freedom.

The corresponding prediction interval including the error variance is

$$\boldsymbol{x}^\top \hat{\boldsymbol{\beta}} \pm s \left(\sqrt{\boldsymbol{x}^\top (\boldsymbol{X}^\top \boldsymbol{X})^{-1} \boldsymbol{x}} + 1 \right) t(n - k, 1 - \alpha/2) \tag{6.10}$$

For the example shown in Figure 6.12 using the training data $\boldsymbol{X}$ we find $\hat{\boldsymbol{\beta}} = (10.4, -17.45)$, $s = 0.483$, $n = 20$, $k = 2$ and

$$(\boldsymbol{X}^\top \boldsymbol{X})^{-1} = \begin{pmatrix} 0.732 & -0.287 \\ -0.287 & 0.162 \end{pmatrix}$$

Evaluating Equation (6.10) for $\boldsymbol{x} = (0, 1)$ the prediction interval is:

$$-17.45 \pm 0.483 \left(\sqrt{0.162} + 1 \right) 2.101 = (-18.875, -16.025)$$

6.3.2 Approximate Prediction Intervals for Nonlinear Models

SR models are usually nonlinear in their parameters and we cannot use the equations above directly. A simple way to determine prediction intervals for nonlinear regression models $Y = f(X, \theta) + Z$ is to use a linear approximation (Bates and Watts, 1988). This is also called the *Delta* method, or – in the context of Bayesian models – Laplace approximation. In this approach we assume that the effects of parameters $\boldsymbol{\theta}$ on the regression function are almost linear in the relevant region around the maximum likelihood estimate $\hat{\boldsymbol{\theta}}$. The method is attractive because it is easy to calculate, but it can lead to bad approximations for the actual prediction interval (Bates and Watts, 1988). However, for large datasets and well-fitting models the likelihood function has narrow peak around the maximum likelihood estimate and the linear approximation is accurate.

To calculate approximate intervals for the expected response we use the equations for linear regression and replace the estimated value $\boldsymbol{x}^\top \hat{\boldsymbol{\beta}}$ by $f(\boldsymbol{x}, \boldsymbol{\theta})$, the matrix $\boldsymbol{X}$ by the Jacobian matrix $\boldsymbol{J}$ evaluated at $\hat{\boldsymbol{\theta}}$, and the vector $\boldsymbol{x}$ by the gradient $\boldsymbol{v}$ with k elements in point $\boldsymbol{x}$.

$$\boldsymbol{v} = \frac{\partial f(\boldsymbol{x}, \boldsymbol{\theta})}{\partial \boldsymbol{\theta}}$$

The $1 - \alpha$ prediction interval including the error variance is

$$f(\boldsymbol{x}, \hat{\boldsymbol{\theta}}) \pm s \left(\sqrt{\boldsymbol{v}^\top (\boldsymbol{J}^\top \boldsymbol{J})^{-1} \boldsymbol{v}} + 1 \right) t(n - k, 1 - \alpha/2) \qquad (6.11)$$

Using the decomposition $\boldsymbol{J} = \boldsymbol{Q}\boldsymbol{R}$, where $\boldsymbol{Q}$ is an orthonormal matrix, $\boldsymbol{R}$ an upper triangular matrix, and $\boldsymbol{R}_1$ the upper p rows of $\boldsymbol{R}$, the expression is simplified to

$$f(\boldsymbol{x}, \hat{\boldsymbol{\theta}}) \pm s \left(\|\boldsymbol{v}^\top \boldsymbol{R}_1^{-1}\|_2 + 1 \right) t(n - k, 1 - \alpha/2) \qquad (6.12)$$

An example where a nonlinear model is appropriate is for the prediction of atmospheric CO_2 content (see Section 7.8). The NOAA ESLR CO_2 dataset is a collection of monthly measurements of relative atmospheric CO_2 content. For this example we only use the data from 2000 until the start of 2020 for training, to find a regression model for CO_2 that only depends on the date. The CO_2 measurements in this data follow a periodic pattern matching the yearly cycle. Therefore, we include *cos(x)* in the function set. The model found by GP is shown in Equation (6.13) with parameter vector $\boldsymbol{\theta} = (2.945, -14.44, -71.72, 16.52, 421.4)$.

$$CO_2 \approx \theta_1 \cos(2\pi\,\mathrm{date} + \theta_2) + \theta_3 \cos(\theta_4 \mathrm{date}^{1/3}) + \theta_5 \qquad (6.13)$$

The model has three linear and two nonlinear parameters (θ_2, θ_4). Using Equation (6.12) we can calculate approximate prediction intervals shown in Figure 6.13. Note that most of the observations are contained in the prediction interval and that the interval stays tight even when extrapolating to 2025. For the fitted model the effect of the nonlinear parameters on the model respose is close to linear in the relevant region around the maximum likelihood estimate and therefore the linear approximation is accurate.

With likelihood profiles (Bates and Watts, 1988) it is possible to calculate more accurate confidence and prediction intervals for nonlinear regression models. Likelihood profiles are intersections of the likelihood function for a parametric model over only a single parameter. To calculate the likelihood profile for one parameter, we change its value in positive and negative direction away from its maximum likelihood estimate in small steps, and in each step reoptimize the values for all other parameters. This provides a profile of the likelihood function over each individual parameter, from which we can determine marginal confidence intervals for each parameter and approximate confidence regions for pairs of parameters. More details can be found in Bates and Watts (1988).

Figure 6.14 shows the approximate pairwise confidence regions for Equation (6.13) which are calculated via the profile likelihood. The pairwise confidence regions deviate only slightly from ellipses. Ellipse-shaped confidence regions are an indicator that the parameters of the model are well estimated and that the peak of the likelihood function is close to Gaussian around the maximum likelihood estimate. Within the small region of high likelihood, the

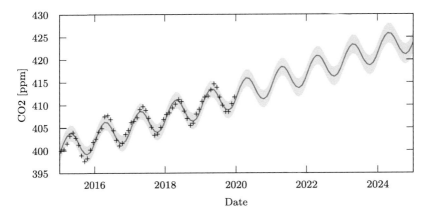

Figure 6.13: Prediction with 95% intervals for the predicted CO_2 fraction for 2015 – 2025.

effect of the nonlinear parameters on the model response is close to linear and the linear approximation is accurate.

In comparison, let us consider another example where the approximate pairwise confidence regions deviate strongly from ellipses. The PCB dataset discussed in Bates and Watts (1988) contains 28 measurements of the concentration of polychlorinated biphenyls (PCBs) in Lake Cayuga trout. The goal is to find a model for the dependency of the PCB concentration and the trout age.

A model found by GP is $\log(\text{conc}) \approx \exp(\theta_1 \, \text{age}) \, \theta_2 + \theta_3$ with $\boldsymbol{\theta} = (-0.19, -3.93, 3.13)$ and $s = 0.497$. The approximate pairwise confidence regions shown in Figure 6.15 indicate that the effect of the coefficients is nonlinear.

Figure 6.16 shows the PCB data and compares the Laplace approximation with the profile-based prediction intervals. Even for this nonlinear model, with large confidence regions, caused by the small number of data points and the high noise level, the Laplace approximation is accurate in the region supported by training data. The prediction intervals calculated via Laplace approximation and likelihood profiles are almost the same and only when extrapolating a small difference becomes visible.

In this book we show prediction intervals for many different SR models and we found that the Laplace approximation works well in almost every case, especially when the SR model fits the data well and the parameter estimates are precise with small marginal confidence intervals. Nevertheless, we recommend to check the pairwise confidence regions for SR models via the profile likelihood calculation to see whether the approximate confidence and prediction intervals are trustworthy.

The Laplace approximation is convenient because it can be calculated easily with almost no additional effort. For example if a given SR implementation

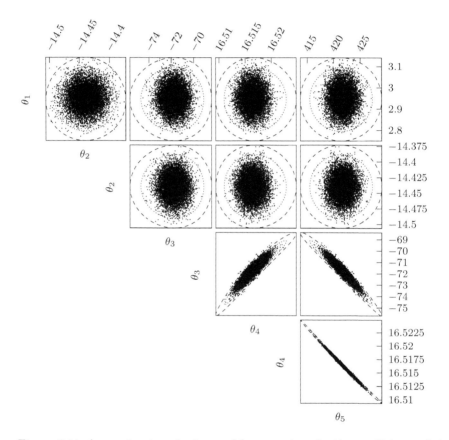

Figure 6.14: Approximate pairwise confidence regions for the coefficients of the CO_2 model Equation (6.13) and MCMC samples from the posterior distribution of the model coefficients for the Bayesian CO_2 model (Equation (6.14)). Both methods produce similar results. $\theta_3, \theta_4, \theta_5$ are correlated and the curved confidence regions for θ_4 show the nonlinear effect of this fitting parameter.

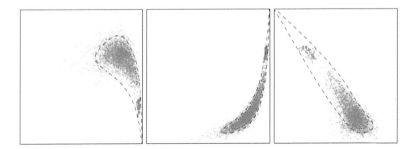

Figure 6.15: The approximate pairwise confidence regions for the coefficients of the PCB model show the strong nonlinearity of the estimated coefficients. The approximate regions match the MCMC samples from the posterior distribution (Equation (6.15)) well.

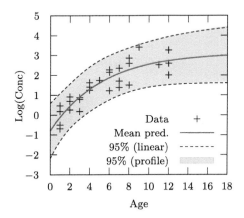

Figure 6.16: The prediction intervals from Laplace approximation and likelihood profiles are almost the same, even though the fitting parameters have a nonlinear effect on the likelihood.

already supports gradient-based local optimization of coefficients then the required functionality for the calculation of confidence and prediction intervals is already available and it is easy to extend the SR implementation accordingly. In contrast, the calculation of the profile likelihood is more difficult to implement correctly and also computationally more expensive because it requires multiple restarts of parameter optimization for the profiles.

6.3.3   Bayesian Prediction Intervals

Alternatively to the methods discussed above, a Bayesian approach allows us to additionally incorporate priors on model parameters and to calculate marginal probabilities for predictions by integrating out model parameters. For Bayesian inference we may use Markov-chain Monte Carlo (MCMC) sampling techniques or different types of approximations. MCMC is a general approach which can be applied with different types of prior distributions and hierarchical models but can be computationally prohibitive. The main difficulty is that MCMC algorithms should find all relevant local optima and additionally sample the region around those modes of the posterior distribution. This is arguably more difficult than finding a single good local optimum as in maximum likelihood esimation.

We use the same two models as above to demonstrate the capabilities of MCMC and compare the results of the profile-likelihood intervals and MCMC. To sample the CO_2 model we use the maximum likelihood estimate $\boldsymbol{\theta}$ and corresponding σ_{err} as a starting point and use the No-U-Turn-Sampler (NUTS) (Hoffman and Gelman, 2014) to generate 10,000 samples. The Bayesian model is

$$\sigma_{\text{err}}^2 \sim \text{truncated}(N(0.8, 1), 0, \infty)$$
$$\boldsymbol{\mu} = (2.945, -14.44, -71.72, 16.52, 421.4)$$
$$\boldsymbol{\sigma} = (7.24 \times 10^{-2}, 2.45 \times 10^{-2}, 1.35, 2.81 \times 10^{-3}, 2.82)$$
$$\boldsymbol{\theta} \sim N(\boldsymbol{\mu}, \text{diag}(\boldsymbol{\sigma}^2))$$
$$\boldsymbol{y} \sim N(CO_2(\text{date}), \sigma_{\text{err}}^2 \boldsymbol{I}) \tag{6.14}$$

where $CO_2(\text{date})$ is the regression function Equation (6.13), and we use a truncated normal prior distribution for σ_{err}. Sampling only takes a few seconds for this model. The samples from the posterior distribution for $\boldsymbol{\theta}$ are visualized in Figure 6.14. The profile-based approximate confidence regions match the empirical distribution of the MCMC samples almost perfectly. Correspondingly, the MCMC prediction intervals also match the profile-based intervals almost exactly as shown in Figure 6.17.

The results are similar for the *PCB* dataset, where we found curved non-

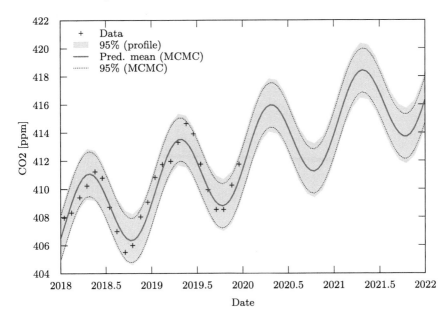

Figure 6.17: Prediction intervals based on MCMC and likelihood profiles match almost exactly for the predicted CO_2 fraction until 2025.

linear confidence regions. Our Bayesian model is

$$\sigma_{\text{err}}^2 \sim \text{truncated}(N(0.25, 1), 0, \infty)$$
$$\boldsymbol{\mu} = (0, 0, 0)$$
$$\boldsymbol{\sigma}^2 = (100, 100, 100)$$
$$\boldsymbol{\theta} \sim N(\boldsymbol{\mu}, \boldsymbol{\sigma}^2)$$
$$\boldsymbol{y} \sim N(\exp(\theta_1 \, \text{age})\, \theta_2 + \theta_3, \sigma_{\text{err}}^2 \, \boldsymbol{I}) \tag{6.15}$$

where we use a rather wide prior for $\boldsymbol{\theta}$ and again use a truncated normal prior for σ_{err} which is based on the maximum likelihood estimate. MCMC sampling of 10,000 points for this simple model with only 28 training points takes only a second or less. Figure 6.15 compares the samples from the posterior parameter distribution with the approximate confidence regions shown above. The curved regions match the MCMC samples surprisingly well even for this dataset with few observations and high noise. The MCMC prediction intervals also match the profile-based intervals as shown in Figure 6.18.

In summary, we found similar results for the Bayesian approach with MCMC sampling and with the profile-likelihood. The simple Laplace approximation works well for the CO_2 model, where parameter estimates are precise as a consequence of a large training set and low noise. MCMC is more powerful than the profile-likelihood but also more demanding.

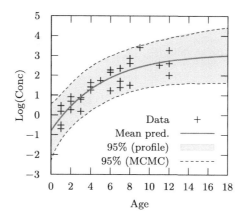

Figure 6.18: MCMC prediction intervals match the profile-based intervals even for this dataset with few training points and high noise.

6.4 Modeling System Dynamics

Differential equations are used to model continuous system dynamics or in other words how a system changes over time. They are fundamental in natural sciences and engineering and it is therefore interesting to consider how such equations can be found automatically from data (Voss et al., 1998, 1999; Schmidt and Lipson, 2009). Before we describe exactly how this can be accomplished with SR, we give a short introduction to differential equations and the necessary terminology.

6.4.1 Basics of Differential Equations

Differential equations are equations that contain derivatives of variables. A simple example is Newton's second law of motion which describes the motion of a mass m when applying a force F.

$$m\frac{d^2x(t)}{dt^2} = F$$

It contains the acceleration which is the second derivative of the position $x(t)$. As a consequence this is called a *second-order* differential equation. In the equation the position x is a function of time as it is variable over time. In fact m and F might also change over time. For conciseness this is usually not stated explicitly in the equation but implicitly assumed.

$$m\frac{d^2x}{dt^2} = F$$

This example is also called an *ordinary differential equation* (ODE) because all derivatives are over a single variable, in this case time t. For ODEs we may also use the more concise dot notation since it is implicit that there is only a single variable over which derivatives are taken.

$$m\ddot{x} = F \tag{6.16}$$

Any second-order equation can be transformed into a system of two *first-order* equations by introducing a new variable.

$$v = \dot{x}$$
$$m\dot{v} = F$$

Equations can be written in *general* and *normal* form. In general form Equation (6.17) one side of the equation is zero. In normal form the left-hand side is one of the derivatives Equation (6.18).

$$0 = m\ddot{x} - F \tag{6.17}$$

$$\ddot{x} = \frac{F}{m} \tag{6.18}$$

An equation in normal form can be easily transformed to an equation in general form. Transforming from the general form to normal form may not always be possible. This is relevant for us because for regression modelling the normal form is assumed and therefore easier to work with when using standard regression tools.

Partial differential equations (PDE) contain derivatives over multiple variables. One example for a second-order partial differential equation is the heat equation which expresses the diffusion of heat u in a medium (here in two spatial dimensions x, y).

$$\frac{\partial u(t, x, y)}{\partial t} = \frac{\partial^2 u(t, x, y)}{\partial x^2} + \frac{\partial^2 u(t, x, y)}{\partial y^2}$$

Again the fact that u is a function of (t, x, y) is implicit and the equation can be written more concisely as

$$\frac{\partial u}{\partial t} = \frac{\partial^2 u}{\partial x^2} + \frac{\partial^2 u}{\partial y^2}$$

Differential equations have an infinite number of solutions whereby each solution fulfills the equation. A solution comes in the form of a function which when substituted into the differential equation allows the equation to be reduced to identity. Consider again Equation (6.16), one possible solution is

$$x(t) = \frac{F}{2m}t^2 \tag{6.19}$$

However, since we lose a constant offset each time we calculate the derivative, all functions

$$x(t) = \frac{F}{2\,m}t^2 + v_0 t + x_0$$

with $v_0, x_0 \in \mathbb{R}$ are solutions to the equation.

Solutions to differential equations such as $\frac{dx}{dy} = f(x, y)$ can be expressed either in *explicit form* $y = F(x)$ or in *implicit form* $G(x, y) = 0$. Solving differential equations does not always lead to an explicit solution in particular when working with nonlinear or higher-order differential equations (Zill and Cullen, 2005). Often the implicit solution is more natural. Again this is relevant because for SR the explicit form is assumed and special techniques are necessary to find implicit solutions (Schmidt and Lipson, 2009).

When solving a differential equation we are usually interested in one particular solution which fulfills additional conditions on the function. For example for Newton's law, we may add the conditions $x(0) = 0, \dot{x}(0) = 0$ to find the solution Equation (6.19). We need two additional conditions to find a solution because it is a second-order differential equation. In this example we specified initial values for the position $x(0)$ and velocity $\dot{x}(0)$. In this case the task of solving the differential equation is called an *initial value problem* (IVP). It is easy to see that alternatively we may also use two different conditions such as the initial position $x(0) = 0$ and the position after one second $x(1) = 10$. In this case the task is a *boundary value problem* (BVP) and in this case leads to the solution

$$x(t) = \frac{F}{2\,m}t^2 + \left(10 - \frac{F}{2\,m}\right)t$$

These two solutions were derived analytically via symbolic manipulation of the differential equation. In general, it can be difficult to solve differential equations analytically and in many cases it is not possible because no closed-form solution may exist. In this case we have the option to solve the equation numerically. This means that instead of finding the function as a mathematical expression we calculate a table of approximate function values for a finite set of function inputs. Through interpolation we can use this approach to get an approximate value for the solution for all elements of the input space. In the examples in the following sections, we always use numerical solutions.

Different approaches are necessary to solve IVP and BVP. Today many good software implementations for numerically solving IVP exist. We will not go into details of these solvers but the general principle is to start with the known initial value and use the differential equation to calculate the change of the variable for a small input step. This is repeated until the whole range of inputs is covered. The simplest form is the Euler method which is a first-order approximation and uses line segments. Much better approximations can however be found using higher order algorithms. The fourth-order Runge-Kutta method (RK4) is a good default which works well and is still relatively easy to implement (see for example Stoer and Bulirsch (2002) and Press et al. (2007)). Solving PDEs is more difficult and is commonly accomplished via discretization and iterative methods. The finite element method (FEM) is an example which uses a mesh-based discretization scheme. In the remainder of the book we discuss only ordinary differential equations.

6.4.2 Finding Differential Equations with Symbolic Regression

There is no fundamental difference when trying to find differential equations from data instead of simple equations. In principle we can use the same techniques that we have already introduced in earlier chapters. However, the nature of differential equations introduces some additional issues that need to be considered, most importantly the amplification of noise in numeric differentiation and the requirement to learn multiple coupled equations for systems of differential equations.

6.4.2.1 Inaccurate Data

For regression modelling we have to select a target variable and input variables from a dataset of measurements of all variables. We usually measure the magnitude of values and not their change over time or space. For example if we observe objects falling from a certain height as in Section 3.2, we would measure position of the object at multiple time points. However, for finding a model in form of a second-order differential equation $\ddot{x} = f(t, x)$ we require data for the second derivative of the measured position. From the measurements we can approximate the velocity, whereby the approximation becomes more accurate by increasing the measurement frequency. This calculation is a numerical approximation of the derivative. In the same way we can approximate the acceleration. Through these steps we can prepare a dataset with multiple observations $(t, x, \dot{x}, \ddot{x})$ and we can then select $\ddot{x}$ as a target variable and all the other variables as inputs and use the direct regression approach.

While this approach of direct SR is natural (cf. (Iba, 2008; Gaucel et al., 2014; Schmidt and Lipson, 2008b)) it may not be successful because of noisy data. The first cause of inaccurate data is the error of the numerical approximation of the derivatives because of limited measurement frequency. The second cause for inaccurate data is that noise in measurements is amplified in each step of numeric approximation of the derivative. As a consequence the approximation of acceleration which is used as a target can be very noisy.

The direct regression approach can still be successful because statistical approaches including SR may find the underlying functional dependency even with high noise levels when a sufficient number of observations is available. However, as a consequence of limited data and limited measurement accuracy we might find that noise level increases so much that it will be impossible to find the functional dependency from the available data.

The first option for improving accuracy is to use an algorithm for the numerical approximation of derivatives which is more robust to noise (Ahnert and Abel, 2007; Breugel et al., 2020). We have found that total variation regularized numerical differentiation (TVDiff) (Chartrand, 2011) works well. TVDiff estimates the finite-difference derivative $\dot{u} \approx \dot{y}$ and promotes smoothness of the estimated derivative by using a loss function which is a weighted combination of the squared error and the total variation penalty

$$L(\boldsymbol{y}, \boldsymbol{u}) = \|\boldsymbol{y} - \boldsymbol{u}\|_2 + \gamma \frac{1}{m} \|\dot{\boldsymbol{u}}_{0:m-1} - \dot{\boldsymbol{u}}_{1:m}\|_1 \qquad (6.20)$$

which can be solved efficiently using a convex solver (Breugel et al., 2020).

Another computationally more efficient option is noise-robust differentiation,[1] and similar techniques for robust numeric differentiation. Figure 6.19 shows the effect for a noisy dataset.[2] The top panel shows the measurements of ψ with a small amount of noise. TVDiff can recover a smooth approximation of $\dot{\psi}$ as shown in the middle panel. In the bottom panel the amplification of noise from taking forward differences is obvious. In fact the noise is so strong that the signal cannot be seen in the bottom panel. Smoothing the results of forward differencing (e.g., using a kernel density smoother) can recover the signal but the smoothed result is subjectively worse than the TVDiff result. It is difficult to get a good approximation of $\ddot{\psi}$ by applying numeric differentiation again.

An alternative approach that can be tried if noise-robust differentiation is not sufficiently accurate, is to skip the approximation of the derivative and instead use a numeric solver to calculate the predicted positions from the differential equation. GP provides a lot of flexibility in how the qualities of evolved expressions are assessed. In the direct regression approach we calculate the SSE of the predicted acceleration to the approximated acceleration derived from the measurements. We may replace this fitness evaluation function by another function which first calls an ODE solver to calculate the predicted positions and then returns the SSE of the measured and predicted positions instead of the accelerations. For initial (or boundary) values we may use measurements from the dataset, or alternatively estimate initial values through fitting, when the measurements are too noisy. This approach does not suffer from the amplification of noise through numeric derivation. However, it is computationally more demanding because we have to solve a differential equation for each solution candidate evaluated by GP. This means that typically we have to call the ODE solver 100,000 to a million times.

This approach of numerically solving the ODE in the fitness evaluation has advantages because it does not use information in the training phase that would not be available in the application. Consider again the example $\dot{y} = f(x, y)$ where we search for an expression for $f(x, y)$ and the causal dependencies between variables shown in Figure 6.20. In the direct regression approach (Iba, 2008; Gaucel et al., 2014; Schmidt and Lipson, 2008b; Quade et al., 2016, 2018,

---

[1]http://www.holoborodko.com/pavel/numerical-methods/numerical-derivative/smooth-low-noise-differentiators/

[2]Taken from https://github.com/SciML/HelicopterSciML.jl

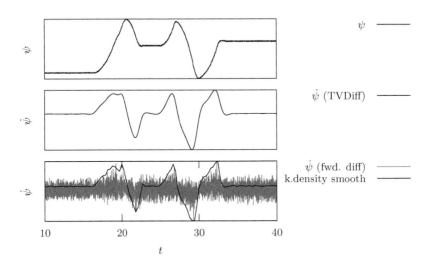

Figure 6.19: Numerical differentiation with TVDiff produces a smooth approximation of $\dot\psi$ (middle). Forward differencing amplifies the small amount of noise in measurements for ψ so strongly that the signal cannot be seen in the visualization anymore. Kernel smoothing of the forward differences produces a worse result than TVDiff.

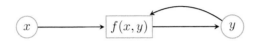

Figure 6.20: Graph of dependencies for $\dot y = f(x, y)$.

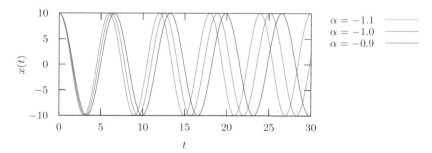

Figure 6.21: Three solutions for the harmonic oscillator $\ddot{x} = \alpha\,x$.

2019), we would approximate $\dot{y}$ numerically and use the observed values of x and y as inputs to the model. This introduces information of $\dot{y}$ in the training process that would not be available in the application of the model. When we use the model for prediction we have to calculate the solution $y(t)$ from the model $\dot{y} = f(x, y, t)$ starting from $y(0)$. This means that the y values we use to assess a model's quality in the training phase may be different from the y values that would result from solving that model. Therefore, the calculated error in the direct regression approach might be a bad estimate of the actual prediction error of the model.

To demonstrate this effect consider a simple oscillator model in Equation (6.21).

$$\ddot{x} = \alpha\,x \qquad\qquad (6.21)$$

The solution for $\alpha = -1$, $x(0) = 10$, $\dot{x}(0) = 0$ is shown in Figure 6.21 together with the solution for $\alpha = -0.9$ and $\alpha = -1.1$. Since the frequencies for the two models differ, the graphs of positions diverge over time resulting in a large SSE. Figure 6.22 shows a plot of the MSE for different parameter values α in $\ddot{x} = \alpha\,x$ assuming normally distributed noise ($\sigma = 0.1$). Correspondingly, the optimal MSE value at $\alpha = -1$ is close to 0.1. We use a direct regression approach, and the reported MSE is for the target $\ddot{x}$ and the model response $\alpha\,x$. When we use the $\ddot{x}$, and x values from the training set, the prediction error only rises slowly when moving away from the optimal α value. However, when we first solve the model and use the numeric solution $x_{\text{pred}}(t)$ as an input ($\ddot{x} = \alpha\,x_{\text{pred}}$), the error function has a much narrower funnel. Correspondingly, α can be estimated more precisely for the same noise level, when we solve the differential equation before calculating errors.

The same argument can be extended to the model structure. Any inaccuracy of the model's parameters or structure is amplified through numerical integration over longer time intervals. Therefore solving the differential equations evolved by GP allows us to recognize and discard inaccurate models much easier.

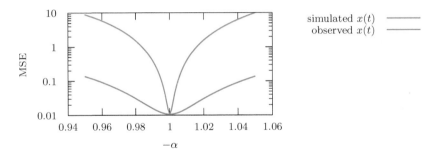

Figure 6.22: MSE for different α-values when using x values from the training set versus the solution of the model x_{pred} as inputs. The error funnel is much narrower when using the numeric solution which means that more precise estimation of α is possible.

6.4.2.2   Finding Systems of Differential Equations

When modelling system dynamics we often have to consider multiple state variables. The corresponding model is a system of differential equations whereby each equation describes the change of one of the variables and may reference all state variables. An example is the SIR Model (Equation (6.22)) to describe the dynamics of infectious diseases in populations. The three state variables in the model represent the proportion of individuals in the population that are susceptible (S) to the disease, infectious (I), and recovered (R).

$$\dot{S}(t) = -\beta\,S(t)\,I(t), \quad \dot{I}(t) = \beta\,S(t)\,I(t) \; - \gamma\,I(t), \quad \dot{R}(t) = \gamma I(t) \quad (6.22)$$

The model is a system of three first-order differential equations in explicit form with two parameters: the infection rate β and the recovery rate γ. The two parameters can be estimated from a time series of observations of the three variables. Note that the equations are coupled; $\dot{I}(t)$ depends on $S(t)$ and $I(t)$. The situation is similar for $\dot{S}(t)$ and $\dot{R}(t)$.

There are several ways to use SR for finding systems of differential equations solely from data. In the approach described by Gaucel et al. (2014), we use SR to find the three equations independently. For this we assume n observations of the time series $S(t_i)$, $I(t_i)$, and $R(t_i)$ at times $t_i, i = 1 \ldots n$ are available, and search for the three expressions for the change rates independently. For efficiency, the derivatives for all three state variables are approximated numerically $\frac{\mathrm{d}Y(t)}{\mathrm{d}t}\big|_{t_i} \approx \frac{\delta Y(t)}{\delta t}$ for all observations, and we fit a separate model to each of the derivative vectors

$$\frac{\delta S}{\delta t} = f_S(S(t_i), I(t_i), R(t_i), t_i, \theta_S) + \epsilon_{S,i}$$

$$\frac{\delta I}{\delta t} = f_I(S(t_i), I(t_i), R(t_i), t_i, \theta_I) + \epsilon_{I,i}$$

$$\frac{\delta R}{\delta t} = f_R(S(t_i), I(t_i), R(t_i), t_i, \theta_R) + \epsilon_{R,i}$$

using three independent SR runs and minimizing $\|\epsilon_S\|_2$, $\|\epsilon_I\|_2$, and $\|\epsilon_R\|_2$. In each run, the observations $S(t_i), I(t_i), R(t_i), t_i, i = 1 \ldots n$ are used as inputs and in the end the three expressions are combined to create the complete model in ODE form

$$\frac{d\hat{S}(t)}{dt} = f_S(\hat{S}(t), \hat{I}(t), \hat{R}(t), t, \theta_S)$$

$$\frac{d\hat{I}(t)}{dt} = f_I(\hat{S}(t), \hat{I}(t), \hat{R}(t), t, \theta_I)$$

$$\frac{d\hat{R}(t)}{dt} = f_R(\hat{S}(t), \hat{I}(t), \hat{R}(t), t, \theta_R)$$

which can be solved numerically for $\hat{S}, \hat{I}, \hat{R}$ using the observed values as starting values $\hat{S}(t_1) := S(t_1)$, $\hat{I}(t_1) := I(t_1)$, $\hat{R}(t_1) := R(t_1)$.

We recommended to execute multiple independent runs and try different combinations of the three expressions to find a system of equations that fits well in combination. This approach of using the approximated derivatives for a regression model is convenient because any regression algorithm can be used and it is efficient. However, as already discussed above, the approach can fail to produce a well-fitting model because the algorithm only fits the model to the approximate derivatives for the observations which can amplify noise in measurements of state variables and increase the chance to find an overfit model. Additionally, the approach searches for the three equations independently and assumes that a perfect model for the other state variables is available. This is implied by using the observed values of all state variables as inputs. However, the solution of the equations $S(t)$, $I(t)$, $R(t)$ may produce completely different predictions. The algorithm only solves the ODE system at the end, and never checks the fit of the solution of the system. The approach can work well, if precise observations (low measurement noise) are available with a high frequency and over the full state space. With higher noise however this approach can be hit or miss.

An alternative is to use a representation that allows to evolve expressions for all equations simultaneously in a singe GP run (Iba, 2008; Kronberger et al., 2020). For this we can use graph-based GP approaches such as Cartesian GP (Miller, 1999) which are especially designed for multiple inputs and multiple outputs (MIMO). In a similar fashion, many tree-based GP systems can be

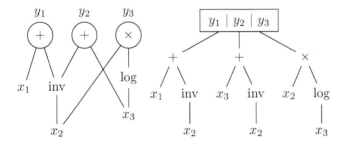

Figure 6.23: Examples for different options to represent SR models composed of multiple expressions.

extended to support multi-tree representations for example by enforcing that the root node has multiple children – one for each expression predicting one of the targets. Figure 6.23 shows examples for graph-based and tree-based SR for multiple target variables. Graph-based approaches have the additional advantage that intermediate results within the graph can be reused within multiple paths to the output nodes which can save computational time.

Regardless of the encoding, it is necessary to carefully define the fitness function for multi-variate SR, especially when the target variables have different scales (e.g., $y_1 \in \{0..0.01\}$ and $y_2 \in \{1000\ldots2000\}$). We recommend to weight the prediction errors for each target variable to guide GP to produce models that are equally accurate for all targets. The sum of the normalized SSE for k target variables

$$\text{fitness}(\boldsymbol{X}, \boldsymbol{\Theta}, \boldsymbol{Y}) = \sum_{i=1}^{k} \frac{1}{\text{var}(\boldsymbol{y}_i)} \|\boldsymbol{y}_i - f_i(\boldsymbol{X}, \boldsymbol{\theta}_i)\|_2^2 \qquad \boldsymbol{\Theta} = \begin{bmatrix} \boldsymbol{\theta}_1^\top, \ldots \boldsymbol{\theta}_k^\top \end{bmatrix} \quad (6.23)$$

where the SSE of each of the the models f_i is taken relative to the variance of each variable $\boldsymbol{y}_i$ is one option. Alternatively one may also use the average over R^2 values which are scale-invariant by design. In this case explicit linear scaling is required before solving the ODE system.

The main benefit from using a multi-tree representation is that the whole system of equations is available for fitness evaluation. It is possible to solve the system of differential equations numerically e.g., using the Runge-Kutta technique to evaluate the goodness-of-fit by comparing the solution to the measurements of $S(t), I(t), R(t)$ directly. On the one hand this approach does not suffer from the amplification of noise in the calculation of numeric derivatives and produces better models because the integration $\hat{S}(t) = S(t_1) + \int f_s(\hat{S}(t), \hat{I}(t), \hat{R}(t), t, \theta_S) dt|_{t_1}^{t}$ is performed for each solution candidate, and the squared error $\|\hat{S}(t_i) - S(t_i)\|_2^2$ can be minimized directly for all state variables. On the other hand this approach requires higher computational effort because each evaluation requires to numerically solve the differential equations. In Chapter 7 we demonstrate this method on several examples.

6.4.2.3   Gradient-based Optimization of Coefficients

As we have discussed in Section 6.2.2 using gradient information for the optimization is essential for the optimization of coefficients. This is also possible for the identification of differential equations. The prerequisite is that the ODE solver supports *sensitivity analysis* which allows to calculate the Jacobian matrix for the solution over the fitting parameters. The Jacobian matrix can then be used for local optimization of the coefficients of the differential equation system in the same way as described in Section 6.2.2.

Using a programming language or software library with support for automatic differentiation it is possible to implement sensitivity analysis by using reverse-mode autodifferentiation (backpropagation) through the solver. A more efficient option is to use adjoint sensitivity analysis (Caracotsios and Stewart, 1985; Cao et al., 2003). This principle has been used for instance for learning differential equations using neural networks (Chen et al., 2018; Brandstetter et al., 2022). The same principles can be applied to optimize the coefficients of SR models.

Regardless of the implementation that is used for gradient-based optimization of coefficients there is the issue of getting stuck in bad local optima. Solutions to differential equations may exhibit many local optima especially for periodic solutions. It can be useful to iteratively optimize cofficients using multiple windows or growing windows from the dataset.

To demonstrate this behaviour we use a Lotka-Volterra model (Equation (6.24))

$$\dot{x} = -0.1\,x + 0.02\,x\,y, \qquad \dot{y} = 0.2\,y - 0.025\,x\,y \qquad (6.24)$$

and simulate the dynamics of this model using a numerical solver for the time interval $t = 0..150$ and $x(0) = 1, y(0) = 2$ to generate a dataset of 151 values $(x(t), y(t))$ at integer time steps. Figure 6.24 shows that local optimization of coefficients can converge to a bad local optimum when the model is fit to the whole dataset. In this case the optimization algorithm converged to a model that predicts the mean target values.

Fitting the differential equation to a shorter time window forces to optimization algorithm to find coefficients which predict short-term dynamics well. This is demonstrated in Figure 6.25 where we use the same initial coefficients as for Figure 6.24. In the first phase the model is fit only to the first 25 points. Then the window is increased to 50 points and coefficients are optimized again using the solution from the previous iteration as a starting point. Iterative retuning with increasingly longer prediction windows allows the optimizer to find coefficients that finally also allow to predict the long-term dynamics well. Initially optimizing the coefficients to shorter windows can improve the solution because the optimizer has to find coefficients that predict short-term dynamics well and can use these coefficients as a starting point to find optimal coefficients for longer predictions.

Additionally, we can use multiple (shorter) windows to make the solution

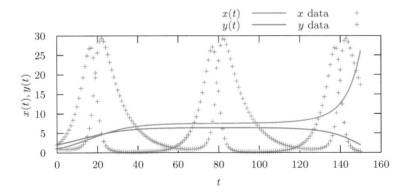

Figure 6.24: Optimization can converge to a bad local optimum when fitting the solution of the ODE to a long time series directly. Here, the solution is close to the average of observed values.

more robust. Instead of increasing the window length we could instead start with n short windows and in each iteration combine two windows to a longer one. This provides more information to the optimization algorithm as it must find a parameterization that works well for different initial conditions.

6.4.2.4 Chaotic Systems

Certain physical systems may exhibit chaotic dynamics, where small variations in the inputs or initial conditions can have large effects in the long term. Examples are the dynamics of the Lorenz model, three-body systems, or the double pendulum. When we try to find differential equations solely from data, we have to compare the simulation of the model to observations. For chaotic dynamics we have no chance to find a model for predicting long-term dynamics, because even if it would be possible to write down the perfect model, small inaccuracies of the model coefficients, initial conditions, or the numerical integration will be amplified quickly and cause the simulation to deviate from the observations. In summary it is not possible to find a model for long-term chaotic dynamics with the same argument that it is not possible to predict chaotic dynamics in the real world. Nevertheless, short-term predictions are possible.

To demonstrate this issue we use the Lorenz equation with three state variables (Equation (6.25)).

$$\dot{x} = a(y - x), \qquad \dot{y} = x(b - z) - y, \qquad \dot{z} = x\,y - c\,z \qquad (6.25)$$

Figure 6.26 shows the solutions for the fourth-order Runge-Kutta (RK4) and Verner's "most efficient" RK9 coefficients[3] (Verner, 1978) (Vern9). For both

---

[3] http://people.math.sfu.ca/~jverner/

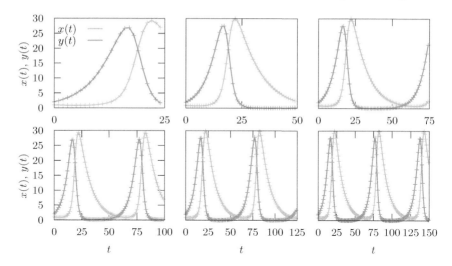

Figure 6.25: Beginning with a small window to optimize the short-term fit and using the solution as a starting point for optimization of the coefficients for larger and larger windows leads to an optimal prediction also for the long term (compare the last panel to Figure 6.24).

solutions we use the exact same model with coefficients $a = 10, b = 28, c = 8/3$ and initial values $x(0) = y(0) = z(0) = 1$. The solutions deviate in the long term because of small differences in the solvers which amplify over time. A similar effect would be observed for minimally different model coefficients or initial values.

However, in Figure 6.26 we also observe that for a limited prediction horizon we would be able to fit a model. It is possible to find the Lorenz model equations with SR if we fit the model to short time windows only.

6.5   Non-numeric Data

So far we have assumed that all data are numeric, most likely real or integer values. This is a requirement so that we can compose SR models from arithmetic operators and real-valued functions such as $\exp(x)$, or $\sin(x)$.

In some applications it may however be necessary to work with non-numeric, qualitative, or categorical variables. This occurs for instance naturally in classification tasks where the target variable is qualitative. However, even in regression tasks it might be necessary to handle qualitative input variables. String values on an ordinal scale (e.g., "freezing", "cold", "warm", "hot", or hardness values using the Rockwell scale) can be mapped to integer values in

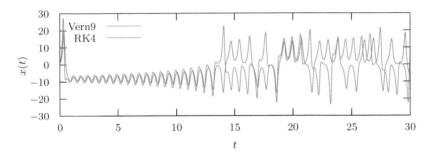

Figure 6.26: Solutions of the Lorenz model (Equation (6.25)) using two different numerical solvers. The solutions differ over the long term because the model has chaotic dynamics and minor differences in solvers or initial conditions can lead to completely different trajectories.

Table 6.3: Example for encoding ordinal and non-ordinal categorical variables.

| Original variables | | Encoded variables | | | |
Hardness	Structure	Hardness	Cubic	Hexagonal	Orthorhombic
B	cubic	2	1	0	0
C	orthorhombic	1	0	0	1
C	cubic	1	1	0	1
A	hexagonal	3	0	1	0

the same order. In this case we rely on SR to identify the potentially nonlinear scale. Values from a non-ordinal scale can be handled through one-hot-encoding whereby a separate binary feature is added for each of the categories. Table 6.3 shows an example of how ordinal and non-ordinal variables can be transformed to numeric variables.

A different approach to include information from categorical variables in SR is to use factor variables which map each of the categorical values to a coefficient (Kronberger et al., 2018). The coefficients for factor variables are fit together with all other coefficients. Factor variables in SR therefore allow different model parameterization depending on the values of categorical variables. This can be useful for applications where we want to find a common model structure which allows representing a family of functions depending on the values of categorical variables. An example where factor variables are used can be found in Section 7.4.

Regression models produce real-valued outputs and cannot predict a categorical variable directly. Therefore, mapping the real-valued response of the model to the categorical target is required.

For target variables with an ordinal scale we may use the same procedure as for input variables and map the string values to integer values while keeping the order relation intact. The integer values can be used for learning the SR

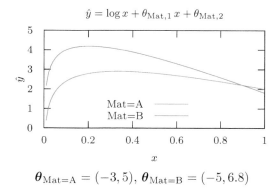

$$\hat{y} = \log x + \theta_{\mathrm{Mat},1}\, x + \theta_{\mathrm{Mat},2}$$

$$\theta_{\mathrm{Mat=A}} = (-3, 5), \ \theta_{\mathrm{Mat=B}} = (-5, 6.8)$$

Figure 6.27: Example of a model with a factor variable (Mat). The model has different coefficients for each value of the factor variable and a common structure to represent the common shape of the function.

model. Additionally, we require a procedure to map the continuous output of the model to the ordered categories. Here we can simply introduce thresholds between the integer-encoded target values and return the corresponding string for each range. This has been discussed briefly in Section 3.4.

6.6 Non-evolutionary Symbolic Regression

The success of metaheuristics like GP is largely owed to their ability to attack different problem domains and perform a general, unconstrained search over the solution space. However, this flexibility does not come for free; the stochastic nature of the evolutionary process makes the validation of results more difficult, as it becomes necessary to compute statistical averages over multiple repeated runs of the algorithm to obtain reliable modelling results. Users have to select from a list of different models with similar goodness-of-fit. As a consequence, the turnover time of modelling results and the time necessary to tune method parameters (e.g., by hyper-parameter search) are increased by a large factor (typically, anywhere from 10x to 50x).

These downsides of stochastic search have led to the development of deterministic approaches that eliminate randomness from the search and impose additional restrictions on the hypothesis space, in some cases greatly limiting the number of possible solution candidates that need to be examined. These determinisic approaches often apply sophisticated heuristics, and search methods (e.g., backtracking, dynamic programming, integration of other ML algorithms). In the following we describe briefly the main concept of a few selected non-evolutionary algorithms for SR.

6.6.1   Fast Function Extraction

Fast Function Extraction (FFX) (McConaghy, 2011) is a "shallow learning" and sparse regression approach to SR that uses regularized learning strategies applied to a large set of generated features. The generation of new features is done by applying several different basis functions B to the original input variables. After the creation of the basis functions, the models are fit using the elastic-net penality (Zou and Hastie, 2005) to promote sparsity. The elastic-net penality is a convex combination of the penality used for ridge regression (Hoerl and Kennard, 1970) and lasso regression (Tibshirani, 1996). The advantages of FFX are that it combines well-established ML techniques to perform SR and scales well with the number of samples and variables in the dataset. Cross-validation is easily applicable for tuning hyperparameters, because FFX is a determinisic algorithm. A drawback of FFX is that it is only able to learn models in the form of generalized linear models (GLMs) (McCullagh and Nelder, 1989), which are specified by the used basis functions. Furthermore, the generated models can become large (≥ 100 basis functions) and although the model is expressed as a linear combination of the basis functions, these large models are difficult to interpret.

6.6.2   Sparse Identification of Nonlinear Dynamics (SINDy)

The algorithm for sparse identification of nonlinear dynamics (SINDy) (Brunton and Kutz, 2022; Quade et al., 2018) is conceptually similar to FFX as it also produces a generalized linear model and uses sparsity promoting techniques to limit the number of terms in the equations. However, while FFX was primarily developed for regression tasks, the main focus of SINDy is to provide similar capabilities for learning models for dynamical systems in the form of ordinary differential equations. SINDy directly supports learning systems of differential equations.

In its basic form, SINDy uses numeric differentiation to calculate derivative values for the observations and then builds regression models to predict those values using the observed values from the training set as inputs. This approach is discussed in more detail in Section 6.4.2.

6.6.3   Prioritized Grammar Enumeration

Prioritized Grammar Enumeration (PGE) (Worm and Chiu, 2013) retains some aspects of GP-based SR but replaces the stochastic elements with dynamic programming and memoization techniques. It constrains the search space to the productions of a predefined grammar, consisting of mathematical expressions and terms. The sentences are stored in a Pareto priority queue (PPQ), generated by non-dominated sorting of all generated sentences according to their accuracy and length. A typical grammar contains rules for adding new terms, widening, or deepening existing terms.

PGE handles the infinite search spaces of possible sentences by reducing them to their canonical form and considering semantics. For example, constant expressions such as $2 + 4$ or $\sin(\pi)$ are automatically folded and terms of commutative symbols (addition and multiplication) are ordered to achieve the canonical form. Therefore, the search space is structured and reduces isomorphic formulas to their simplest representation. The formulas already evaluated by PGE are kept in a trie structure to avoid reevaluation of equivalent equations, hence saving execution time. The method is deterministic and does not have to be executed multiple times. A drawback of PGE is that with an increasing number of variables its runtime increases drastically, because the search space grows exponentially with the number of variables.

6.6.4   Exhaustive Symbolic Regression

Exhaustive Symbolic Regression (ESR)[4] (Bartlett et al., 2023b) is a deterministic algorithm that considers the set of all possible equations that can be generated from a given set of basis operations up to a specified maximum complexity. If the search space defined in this way contains the global optimum, then the corresponding expression is guaranteed to be discovered. ESR employs the Minimum Description Length principle (Hansen and Yu, 2001) to ensure that the selected models represent the best compromise between accuracy and simplicity.

Expressions are represented as trees which are encoded as lists of nodes in prefix (depth-first) order. Each tree is generated starting from a single root node until no more children can be added to any of the nodes. The accepted operators can be binary, unary, or nullary (parameters or variables). The ESR algorithm incorporates logic for detecting duplicate expressions (trees that represent the same function) and is thus able to handle an exponentially-growing number of equations. This is accomplished by detecting simplifications, reparameterisation invariance, or parameter combinations (in which case, expressions are considered equivalent).

ESR is implemented in the Python programming language and makes use of the Sympy library (Meurer et al., 2017) for performing the necessary algebraic manipulations. The generated function sets up to a given complexity can be stored for later use to save time on the generation step in further applications. The approach obtains good results and outperforms similar approaches in modelling cosmological data. It is best suited for finding functions of one, two, or maximally a handful of input variables because the size of the ESR search space grows exponentially with the number of variables. An implementation of the algorithm can be found at `https://github.com/DeaglanBartlett/ESR`.

---

[4]Open source implementation: `https://github.com/DeaglanBartlett/ESR`

6.6.5 Grammar-guided Exhaustive Equation Search

Similar to PGE and ESR, the method described by Kammerer et al. (2020) also exhaustively evaluates all equations in the search space, but allows to restrict the structure of equations with a context-free formal grammar. As discussed in Chapter 4, grammars make it easy to include domain knowledge. Examples include fixing parts of the equations, providing building blocks – certain combinations of functions, operators, and even terminal symbols – that should occur in the equations, and limiting the complexity of equations.

While ESR produces all possible unique equations in parallel, grammar-guided equation search constructs equations iteratively starting with the shortest equations. The process simply uses the grammar rules to expand phrases following a heuristic search in the directed acyclic graph of phrases that are possible to generate from the grammar. The root of the graph is the sentence symbol of the grammar from which all equations can be derived. Semantic duplicates for example because of commutative operators are detected through semantic hashing to prevent revisiting all equations that are extensions of such duplicates. This reduces the size of the relevant parts of the search graph, especially for problem instances with many variables. The constructive approach in combination with the detection of semantic duplicates allows to evaluate only unique equations without generating all equations first, which is prohibitive for problem instances with more than a few variables.

Alternatively to the simple breath-first search, which allows to search the whole search space exhaustively, heuristic search can be used to search more intensively in the part of the search graph which contains extensions of the best equation found so far. An implementation of the algorithm in C# can be found at `https://github.com/heal-research/HEAL.EquationSearch`.

6.6.6 Equation Learner

Equation Learner (Sahoo et al., 2018) employs a shallow neural architecture called an "equation learning network" (EQL) to learn the expressions governing the behaviour of real-world physical systems. The EQL is a fully connected feed-forward neural network whose nodes include base functions commonly used in symbolic equations (e.g., $+$, $\times$, *sin*, *cos*, etc.). Compositions of these base functions are generated across the network layers.

Regularization phases originally proposed by Martius and Lampert (2017) are used to sparsify the network connections to reduce the number of terms to allow extracting a expression resembling a typical formula describing a physical system. Sparsity is considered as a criterion for model selection alongside validation error.

One challenging aspect in this type of architecture is the inclusion of division symbols, which by default introduce poles with abrupt changes in convexity and diverging function values which cause issues for backpropagation learning. The authors sidestep this issue by using the assumption that any real system cannot generate data at the pole itself and assume that a single branch of the hyperbola $1/b$ with $b > 0$ suffices as a basis function and furthermore, divisions are only allowed in the output layer. To increase robustness, the usual division is replaced with a regularized version in which an arbitrary value $\theta \geq 0$ is used as an activation threshold, and a penalty term is used to steer the network away from negative values in the denominator.

6.6.7   Deep Symbolic Regression

Deep symbolic regression (DeepSR) (Petersen et al., 2021) uses reinforcement learning to train a deep recurrent neural network (RNN) to generate symbolic expression trees. The reward function is inspired from SR and uses a "squashed" expression of the normalized mean squared error that results in a more stable gradient. A reward baseline as well as a complexity penalty are additionally introduced to reduce variance and to bias the tree generation towards simplicity. An entropy bonus is introduced to prevent premature convergence and increase the diversity of the generated expressions. A risk-seeking policy is used to improve best-case performance by using only the top ϵ percentile samples from each batch for gradient computation.

Each tree is generated one node at a time along a preorder traversal of the partial tree structure, with the parent and sibling of the next node being used as the next input to the RNN. The method allows constraints over the search space defined as limits on the minimum and maximum tree length or restrictions on possible parent-child node combinations (e.g., the child nodes should not be all constants, the child of an unary operator should not be the inverse of that operator, no nesting of trigonometric operators, etc.). These rules are applied concurrently with the autoregressive sampling by zeroing out the probabilities of selecting node symbols that would violate any constraint. Once a preorder traversal is sampled, a nonlinear least squares optimization step is used to optimize the coefficients of the corresponding symbolic expression.

6.6.8   AI Feynman

AI Feynman (Udrescu et al., 2020) is a hybrid approach combining conventional solving approaches with neural networks. It uses a divide-and-conquer approach to find exact solutions to regression problems, mainly from the physical domain, by recursively detecting and exploiting modularity in the function's computational graph. The key assumption is that the sought function can be represented as a composition of a basis set S of simpler functions, and that this representation can be specified as a graph whose nodes contain elements of S.

If the problem cannot be solved by polynomial fitting or brute force search, then the method trains a neural network and uses it to estimate functional modularities, which are then used to separate the problem into smaller and simpler sub-problems. The software then recursively applies various solving strategies to discover parts of the computational graph and is able to recognize properties like compositionality, symmetry or generalized additivity. In order to decrease sensitivity to noise and outliers, avoid overfitting and generate simpler solutions, the model description length is minimized alongside training error.

Discovered graph modules are then merged together in a bottom-up fashion, until the structure of the entire graph is reverse-engineered. A set of optimizations are employed to improve the speed of the algorithm: Pareto-optimal pruning, greedy search of simplification options, or accelerated brute search with the help of hypothesis testing.

7

Examples and Applications

Many domains in science and engineering offer examples for SR modelling tasks. In this chapter we discuss several examples in detail whereby we try to cover many different domains. Some of the examples demonstrate advanced techniques that we have described in the earlier chapters. For each example we give clear instructions how to use SR and describe the full modelling process including data collection, preprocessing, algorithm parameterization, model selection, simplification, interpretation, and evaluation. These examples should enable you to successfully apply SR for your particular application.

7.1 Yacht Hydrodynamics

With the computing power available today, physics-based aerodynamic and hydrodynamic simulation models allow extremely detailed and accurate simulation of the behaviour of wing shapes and are quintessential for design and optimization of bodies moving through gases or fluids. These high-fidelity simulations can be computationally demanding, putting a limit to the number of different designs that can be simulated even when using high-performance computing facilities. In such cases it can be interesting to build simplified simulation models (surrogate models) that can be used to explore the space of potential designs more rapidly and reduce the set of possible designs for which the high-fidelity model has to be run. To generate such simplified models we can use ML methods including SR whereby we have mainly two options to gather data for training. Either we collect data through measurements from the physical system itself, or we generate data using the high-fidelity simulation model. Most of the time the natural choice is to use data from the high-fidelity model, because measurement would either be too costly or not possible at all. Measurement requires the system to physically exist for instance in the form of different prototypes (physical models) that have to be built. Simulation is more convenient but requires an accurate physics-based simulation model.

In the following we focus on the prediction of the resistance of different yacht hull shapes at different speeds using a well-known classic dataset. The goal is to produce a prediction model that can be used to find an optimal shape for a given speed range.

The *yacht* dataset contains measurements for the resistance of different yacht shapes and has been collected from prototypes (Gerritsma et al., 1981). The dataset has been published already in 1981 and is sometimes used for benchmarking ML methods even today.

The data collection process involved building 22 different hull form models and a total of 308 experiments (22 models at 14 speeds) in tanks of the Delft Ship Hydromechanics Laboratory (Gerritsma et al., 1981). Gerritsma et al. used six dimensionless relative parameters that were calculated from absolute lengths. This parameterization is very convenient for modelling as it removes almost all dependencies between parameters and automatically allows to scale the mathematical model to longer yachts (see Section 6.1.5.3). The 22 tested hull shapes cover the design space well (see Figure 7.1).

The absolute and relative hull form parameters are given in Table 7.1. For this demonstration we only use the relative parameters. The residuary resistance, expressed via the Froude number F_n, is the resistance per unit weight of displacement and mainly depends on the velocity. The Froude number $F_n = V/\sqrt{g\,\mathrm{LWL}}$ for velocity V and gravitational force g is a dimensionless parameter that can be used in this context to quantify velocity. To generate the dataset for SR we use LCB, C_p, $\frac{\mathrm{LWL}}{\nabla_c^{1/3}}$, $\frac{\mathrm{BWL}}{T_c}$, $\frac{\mathrm{LWL}}{\mathrm{BWL}}$, and F_n as input variables and the residuary resistance as the target. Using this parameterization we will produce a single model for all F_n values. This is in contrast to the polynomial model used in (Gerritsma et al., 1981) which does not include F_n. Instead Gerritsma et al. decided to split the dataset by F_n and fit the polynomial model for each of the velocities separately, resulting in 14 different coefficient vectors.

Table 7.1: Absolute and relative hull form parameters (Gerritsma et al., 1981). We use only the relative parameters in the second group for SR.

Parameter	Description
LWL	Waterline length
BWL	Waterline beam (width)
B_{max}	Maximum beam
T_c	Draught of canoe body
	(max. vertical distance from waterline to bottom of canoe)
∇_c	Volume of displacement of canoe body
LCB (%)	Relative longitudinal position of the center of buoyancy
C_p	Prismatic coefficient
$\frac{\mathrm{LWL}}{\nabla_c^{1/3}}$	Length to displacement ratio
$\frac{\mathrm{BWL}}{T_c}$	Beam to draught ratio
$\frac{\mathrm{LWL}}{\mathrm{BWL}}$	Length to beam ratio

The scatter plot matrix in Figure 7.1 shows the strong nonlinear dependency of residuary resistance and F_n and the good coverage of the design space as none of the pairwise plots of the hull parameters show a strong correlation. The main challenge in this task is to model the small effects of the hull parameters which cause the small deviations in the resistance for each value of F_n.

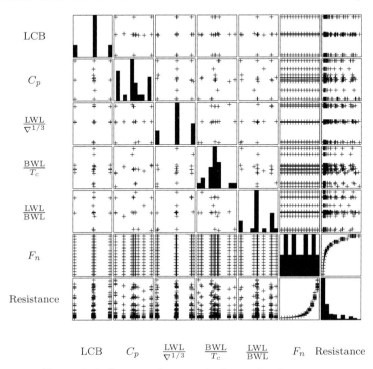

$$LCB \qquad C_p \qquad \frac{LWL}{\nabla^{1/3}} \qquad \frac{BWL}{T_c} \qquad \frac{LWL}{BWL} \qquad F_n \quad Resistance$$

Figure 7.1: Scatter plot matrix for yacht dataset.

First, we have to determine GP parameters that work well for this problem. As we do not know anything about the required complexity and runtime for this problem, we try several different function sets and limits for the maximum size of trees. The function sets are F_{Poly} for polynomial models, F_{RatPoly} for rational polynomials, and F_{General} which allows general expressions.

The maximum size is varied from very small models with ten nodes to larger models with 100 nodes. Concretely, we use the values $10, 15, 20, 25, 30, 35, 40, 45, 50, 75, 100$ with larger step size for the larger values because the runs with larger size limits require longer run times and small relative changes in the size limits likely only have a small effect on the achievable goodness-of-fit. The remaining GP parameter values are shown in Table 7.2.

As the dataset is relatively small, we use systematic cross-validation to get a better estimate of the expected generalization error for the models produced with each configuration. For this we systematically split the dataset into 11 folds, whereby each fold contains all measurements for two hull forms. This

Table 7.2: Algorithm parameters for the yacht hydrodynamics problem.

Parameter	Value
Population size	1000
Generations	50
Objective	MSE (minimize)
Coefficient optimization	10 Levenberg-Marquardt iterations
Selection	Tournament (group size=5)
Crossover rate	100%
Mutation rate	25%
Replacement	Generational (1 elite)
Max. tree depth	10 levels
Max. tree sizes	$\{10, 15, 20, 25, 30, 35, 40, 45, 50, 75, 100\}$ nodes
Function sets	$F_{\text{Poly}} = \{+, \times\}$
	$F_{\text{RatPoly}} = \{+, \times, \div\}$
	$F_{\text{General}} = \{+, \times, \div, \exp(x), \log(x), x^2, \sqrt{x}\}$

systematic assignment of measurements to folds guarantees that we test each model on data for two new hull shapes that are not used for training. The i-th hull shape ($i = 1 \ldots 22$) is assigned to fold $(i - 1) \bmod 11$. For each fold we execute ten independent GP runs using the same parameterization but different random seeds to improve the estimate of the expected generalization error. From the same runs we can also determine the number of generations that is required to reach a good solution. This can be accomplished by logging the NMSE for the training and test partition for the best solution in each generation. This allows us to calculate the average NMSE values for each configuration over generations and we can select the smallest generation where the average test NMSE reaches a minimum.

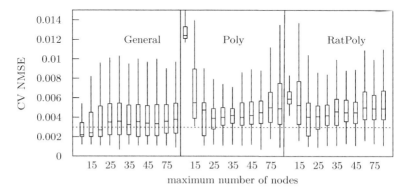

Figure 7.2: Box plot for the NMSE values on the test folds from cross-validation. The dashed horizontal marks the mean NMSE of the best configuration (Function set F_{General} with ten nodes).

Figure 7.2 shows box plots of the NMSE on the test folds for each configuration (11×10 results for each configuration). The minimum mean NMSE of the best configuration (F_{General} with maximum tree size 10) is 0.00298 and visualized as a horizontal dashed line. This minimum mean NMSE value is reached after only eight generations for this configuration. The results for the best configuration have a small variance and all other configurations have higher median values than the best configuration.

Above we set the number of generations arbitrarily to a rather small value of fifty. Prior experience shows that with memetic local optimization of numeric parameters a small number of twenty to 100 generations is often sufficient to find acceptable solutions. For this problem we need even fewer generations to produce a good model.

Before we start the final GP run it is worthwhile to check the quality line charts of the selected configuration. Figure 7.3 shows the NMSE values on training (yellow) and test sets (blue) for the best solution over generations for all 11 folds in the left panel. The NMSE values have high variance caused by the small test set. More importantly however, the average test error increases from generation eight onwards. Therefore, we select the configuration F_{General} with maximum tree size ten and eight generations for the final run.

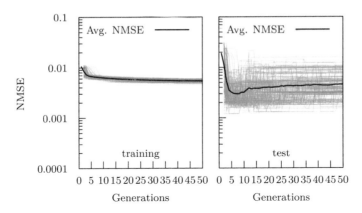

Figure 7.3: Line chart of the NMSE for the training and test sets. The thick line marks the average test NMSE over generations. After seven generations the algorithm starts to overfit.

Figure 7.4 shows scatter plots of measured and predicted residuary resistance for the training best SR model produced in our final run. The prediction for the measurements for the two hull models in the test set have similar deviations as for the models in the training set. Figure 7.5 shows the tree and the corresponding expression. The model has only four coefficients and uses only three of the variables whereby an interaction between C_p and LCB and C_p and F_n has been identified.

As we have seen before, the model response plot shown in Figure 7.6 can

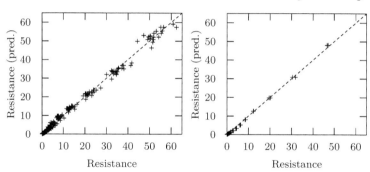

Figure 7.4: Scatter plots for the measured and predicted residuary resistance for training and test partitions for the first cross-validation fold.

help immensely for the interpretation of the model. As stated above the main effect is F_n and we are mainly interested in the smaller deviations in resistance caused by the hull parameters. Therefore, we show all response curves produced for the different F_n values and align the curves by plotting the deviation from the resistance value that is predicted for average hull parameter values on the y-axis. The plots of the model response show interactions of hull parameters with the velocity as expressed through F_n. As the longitudinal position of the center of buoyancy (LCB) is increased, the resistance also increased whereby the effect is stronger for higher velocities. A small LCB means that the center of buoyancy is to the back of the canoe. A higher prismatic coefficient (C_p) decreases resistance whereby again the effect is minimal for small velocities.

$$(0.251 - 0.186\,C_p)^2 \left(0.315\,\mathrm{LCB} + e^{8.716\,F_n}\right)^2 \qquad (7.1)$$

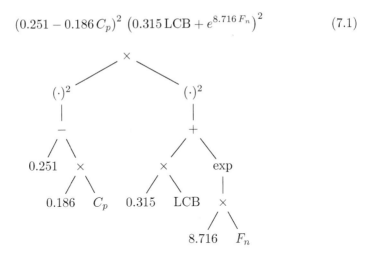

Figure 7.5: Expression for predicting the residuary resistance and the corresponding tree evolved by GP.

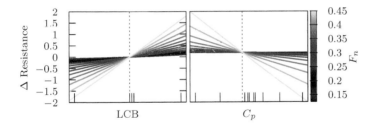

Figure 7.6: Model response plots for the yacht model. The plots show the predicted *deviation* in resistance for F_n values (color coded) over the center of buoyancy (LCB) and the prismatic coefficient (C_p).

7.2 Industrial Chemical Processes

The *Tower* and *Chemical* datasets originate from industrial processes at the Dow Chemical company (Kordon, 2008), and have been used in several articles to test GP for SR.

The *Chemical* dataset contains 58 variables and 1066 observations. The target variable represents noisy measurements of the chemical composition of an end-product, while the remaining variables represent input features including process parameters such as material flows, pressures and temperatures.

The *Tower* dataset contains 26 variables and 4999 observations. The target variable represents the propylene concentration at the top of a distillation tower, measured regularly every 15 minutes using gas chromatography. The rest of the variables represent the process parameters and are measured every minute. To synchronize with the target variable, the remaining variables are aggregated into 15 minute averages (Vladislavleva et al., 2009).

Our modelling approach consists of two main steps: a data preprocessing step where we perform correlation analysis and apply a basic form of feature selection to the data, and a modelling step where we apply SR using both original and preprocessed datasets. For each of these datasets, two-thirds of the observations will be used for training while the remaining one-third will be used as test data.

7.2.1 Correlation Analysis

Correlation analysis represents a first preprocessing step in the modelling process, aimed at eliminating redundancy in the data by identifying highly correlated input variables. For this purpose, pairwise correlation coefficients (ρ) between all input variables are calculated. As a rule of thumb, input features with a ρ^2 value of at least 0.95 can be considered to be highly correlated. As such, one of the two features can be excluded.

On the *Chemical* dataset we notice three groups of correlated variables, as illustrated graphically in Figure 7.7. Therefore, a number of eight input features can be excluded:

- Variables $\mathbf{x_{17}}$, x_{18}, x_{19}

- Variables $\mathbf{x_{23}}$, x_{24}, x_{26}, x_{27}, x_{28}

- Variables $\mathbf{x_{33}}$, x_{52}, x_{53}

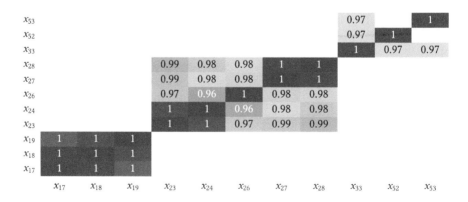

Figure 7.7: *Chemical* dataset correlation matrix. Input features with squared correlation ρ^2 below 0.95 are omitted.

Based on this information, considering set S to be the set of all independent variables, we can define set

$$S' = S \setminus \{x_{18}, x_{19}, x_{24}, x_{26}, x_{27}, x_{28}, x_{52}, x_{53}\}$$

as our filtered input set where correlated variables are excluded. Note in this case the variables shown in bold font above are kept as a representative from each group.

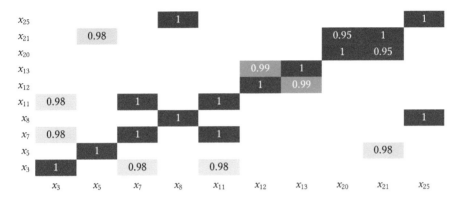

Figure 7.8: *Tower* dataset correlation matrix. Input features with squared correlation ρ^2 below 0.95 are omitted.

For the *Tower* dataset, correlation analysis (shown in Figure 7.8) reveals a few groups of correlated variables:

- **x_3**, x_7, x_{11}

- **x_5**, x_{21}

- **x_{20}**, x_{21}

- **x_8**, x_{25}

- **x_{12}**, x_{13}

We note here that although both x_5 and x_{20} are highly correlated with x_{21}, they are not highly correlated ($\rho^2 > 0.95$) with each other, therefore the cautious approach is to keep both as input features. In this case we obtain input subset

$$S' = S \setminus \{x_7, x_{11}, x_{13}, x_{21}, x_{25}\}$$

where S represents the set of all 25 input variables.

7.2.2 Experimental Setup

As SR is generally implemented as a stochastic approach, each tested configuration must be repeated a sufficient number of times to ensure the statistical stability of the results. Therefore, we repeat the modelling process fifty times for each problem. At the end we present the results as aggregated median performance measures and then extract the best performing model on the training set for each problem.

We configure the algorithm to use a population size of 1000 individuals and 500 generations. Parents are selected using a tournament with group size five, crossover rate is 100%, random mutation is performed for 25% of the solution candidates produced via crossover, fitness function is the squared correlation coefficient ρ^2 without optimization of coefficients. The function set is $\mathscr{F} = \{+, -, \times, \div, \exp, \log\}$, and randomly initialized coefficients as well as variables weighted with a coefficient are used as terminal symbols. The maximum size of trees is limited to fifty nodes and 12 levels.

7.2.3   Results for the Chemical Dataset

We perform two experiments for the sets S and S', to analyze the benefits of feature correlation analysis in the preprocessing step. Although SR performs implicit feature selection within its search loop, it is beneficial to eliminate highly correlated variables from the start to ease the search and find more similar models.

We use 2/3 of the dataset for training and the remaining 1/3 for testing models, resulting in 711 training samples and 355 test samples.

It can be seen from the results shown in Table 7.3 that the two configurations perform quite similar to each other. Indeed, statistical testing using the Kruskal test (Kruskal and Wallis, 1952) reveals no significant difference between the two configurations. In the postprocessing step, we select the best performing model on the training set from each configuration and manually simplify its structure, by a) performing arithmetic simplifications, and b) pruning parts of the tree structure if they have a negative or near-zero impact on its output. After this step is complete, we optimize the numerical coefficients of the simplified model using nonlinear least squares.

Table 7.3: Aggregate results for the *Chemical* problem. Training and test performance shown for each terminal set as median $R^2 \pm$ standard deviation.

Chemical problem	R^2 train	R^2 test
Full input set S	0.782 ± 0.016	0.733 ± 0.194
Reduced input subset S'	0.775 ± 0.015	0.744 ± 0.118

The best training solution obtained with the full input set S (Equation (7.2)), contains 12 input variables and has a training $R^2 = 0.81$ but much worse test $R^2 = 0.56$. By selecting the best model on the training set, we selected an overfit model with worse than average test R^2.

$$\hat{y}_S = \left(\frac{(\theta_1 \, x_7 + \theta_2 \, x_{33}) \, x_{12} \, x_{56} + \theta_3 \, x_{56}}{\theta_4 \, x_{35} + x_5} \right.$$

$$\left. + \, (\theta_5 \, x_7 + \theta_6 \, x_{22} + \theta_7 \, x_{42} + \theta_8 \, x_{47}) \, x_{49} \right) \frac{x_{17}}{x_{16}} + \theta_9 \qquad (7.2)$$

The best solution obtained when using subset S' (Equation (7.3)), also has 12 input variables and achieves a training $R^2 = 0.821$ and test $R^2 = 0.83$. Here we were lucky that the selected model also has an above average test R^2.

$$\hat{y}_{S'} = \left(\theta_1 \, x_{29} + \theta_2 \, x_{45} + \theta_3 \, x_{46} + \theta_4 \, x_{47} + \frac{x_{49}}{x_5 + \theta_5} \right) \frac{\theta_6 \, x_{16} + \theta_7 \, x_{49}}{\log x_7}$$

$$+ \, \theta_8 \, x_{36} + \theta_9 \, x_{51} + \theta_{10} \, x_{55} \, x_{57} + \theta_{11} \qquad (7.3)$$

As shown in Table 7.4, the two models share a common subset of variables: x_5, x_7, x_{16}, x_{47}, and x_{49} but the importances in the individual models as expressed by the SSR ratio (Equation (5.4)) differ. The variable frequencies shown in Table 7.5 averaged over all fifty runs for both configuration produce a similar ordering for both configurations, which demonstrates GP's ability to perform feature selection during the search. The top three variables ordered by frequency are the same for both configurations.

None of the correlated features occurs in the top ten for the runs with the full feature set S. This implies that the groups of correlated features are not important for producing a well-fitting prediction model and explains why there is no difference in the GP results for both configurations. In conclusion the removal of correlated variables did not prove beneficial for the *Chemical* dataset.

7.2.4 Results for the Tower Dataset

The *Tower* data containing 4999 samples is split into 3136 training samples and 1863 test samples. We again perform fifty repetitions for the two configurations using the full input set S and filtered subset S'.

The median results obtained by the two configurations are displayed in Table 7.6, again showing almost no difference between the two configurations. The best solution using the full input set S (Equation (7.4)) contains six variables and has $R^2 = 0.886$ on the training set and the same R^2 on the test set.

$$\hat{y}_S = \left(\frac{\theta_1 \, x_6}{1 + \theta_3 \, x_{15}} + \frac{\theta_4 \, x_6}{\log x_1 + \theta_5} + \theta_6 \, \frac{x_2}{x_1} \right.$$

$$\left. + \, \theta_7 \, \log(x_1) \, x_{15} + \theta_8 \, x_{15} + \theta_9 \, x_{12} + \theta_{10} \, x_{23} \right)^{-1} + \theta_{11} \qquad (7.4)$$

Table 7.4: Model-specific variable importance scores for model $\hat{y}_S$ (left) and $\hat{y}_{S'}$ (right) on the *Chemical* dataset.

Variable	SSR ratio		Variable	SSR ratio
x_{47}	2.52		x_{49}	3.46
x_{16}	2.36		x_{47}	2.33
x_7	2.02		x_7	1.95
x_{17}	1.93		x_5	1.60
x_{49}	1.81		x_{16}	1.41
x_5	1.67		x_{29}	1.28
x_{42}	1.23		x_{46}	1.19
x_{33}	1.16		x_{45}	1.19
x_{12}	1.05		x_{36}	1.13
x_{56}	1.03		x_{51}	1.10
x_{35}	1.01		x_{57}	1.08
x_{22}	1.01		x_{55}	1.03

The best solution using the input subset S' (Equation (7.5)) has nine input variables and achieves a better R^2 score of 0.905 on the training set and the same R^2 on the test set.

$$\hat{y}_{S'} = \frac{\theta_1}{\theta_2\, x_1 + \theta_3\, x_6 + 1} + \frac{\theta_4\, \log x_{10} + \theta_5}{\log x_2 + \theta_6} + \frac{\theta_7\, x_2}{x_1 + \theta_8} + \theta_9\, x_1 + \theta_{10}\, x_6 + \theta_{11}\, x_8$$
$$+ \theta_{12}\, x_{12} + \theta_{13}\, x_{14} + \theta_{14}\, x_{15} + \theta_{15}\, x_{23} + \theta_{16} \tag{7.5}$$

All seven variables $(x_1, x_2, x_6, x_{12}, x_{15}, x_{23})$ contained in the model $\hat{y}_S$ also appear in the model $\hat{y}_{S'}$. Noteably, the correlated variables which were removed for S' are not used in the best model trained for the full feature set, which again implies that the identified groups of correlated variables are less important for predicting the target variable.

The best model for the reduced feature set $\hat{y}_{S'}$ uses three additional variables (x_8, x_{10}, x_{14}) but they only have low impact in the model as shown in Table 7.7. In both models variables x_1, x_6, x_{12}, x_{23} have the highest impact.

This outcome demonstrates the ability of the evolutionary process to implicitly select the best features to be included in the model. Despite starting with two different input feature sets, the two models ended up including almost the same variables. Furthermore, the variables included in the models obtain similar rankings in the two models y_S and $y_{S'}$ (see Table 7.7). The average variable frequencies over all fifty GP runs for both configurations shown in Table 7.8 are again similar for the full and the reduced variable sets. x_6, x_1, x_{23}, x_4 occur most frequently on average for both configurations. When using the full variable set, both x_{12} and x_{13} which are highly correlated occur frequently, none of the other excluded variables occur in the top 10.

Table 7.5: Top 10 variables ordered variable frequency (averaged over the fifty GP runs) for both configurations for the *Chemical* dataset. Left: full feature set S, right: reduced feature set S'.

Variable	Avg. freq. [%]	Variable	Avg. freq. [%]
x_{49}	18.3	x_{49}	18.7
x_{47}	10	x_{47}	11.9
x_{16}	7.6	x_{16}	7.6
x_1	6.4	x_{54}	5.3
x_{29}	5.3	x_1	4.9
x_{54}	5.3	x_{29}	4.9
x_{35}	4.1	x_{35}	4.6
x_{42}	2.7	x_{38}	2.6
x_{38}	2.5	x_{36}	2.4
x_{46}	2.1	x_{42}	2.3

Table 7.6: Aggregate results for the *Tower* problem. Training and test performance shown for each input subset as median $R^2 \pm$ standard deviation.

Tower problem	R^2 train	R^2 test
Full input set S	0.869 ± 0.019	0.868 ± 0.024
Input subset S'	0.865 ± 0.016	0.863 ± 0.017

Table 7.7: Variable importance scores for model $\hat{y}_S$ (left) and $\hat{y}_{S'}$ (right) on the *Tower* dataset.

Variable	SSR ratio	Variable	SSR ratio
x_6	31.74	x_1	2.51
x_{12}	6.65	x_6	2.45
x_1	4.38	x_{23}	1.92
x_{23}	3.46	x_{12}	1.53
x_{15}	1.43	x_2	1.21
x_2	1.05	x_{14}	1.11
		x_{10}	1.07
		x_{15}	1.07
		x_8	1.05

Table 7.8: Top 10 variables ordered variable frequency (averaged over the fifty GP runs) for both configurations for the *Tower* dataset. Left: full feature set S, right: reduced feature set S'.

Variable	Avg. freq. [%]	Variable	Avg. freq. [%]
x_6	16.4	x_6	19.6
x_1	15.8	x_1	15.3
x_{23}	14.6	x_{23}	13.0
x_4	7.7	x_4	8.9
x_{15}	5.3	x_2	2.9
x_2	3.7	x_3	3.2
x_9	4.7	x_{15}	4.2
x_{13}	4.1	x_9	5.0
x_{12}	4.9	x_8	2.5
x_3	2.6	x_{12}	5.1

7.3 Interatomic Potentials

This example demonstrates the use of SR to determine the Potential Energy Surface (PES) of an atomic system. The PES is a conceptual tool that illustrates the relationship between the geometry of a set of atoms and its potential energy. It has important applications in the analysis of chemical reaction dynamics, transition states, chemical stability, and other material properties in materials science (Hernandez et al., 2019).

In molecular dynamics simulations, the total potential energy of the system is typically obtained as a sum of local energies resulting from m-body particle interactions, which are modeled by functions of each particle's position, $\mathbf{r}$.

The simplest way of calculating the total potential energy of the system is to assume that total energy is the sum of the energy contributions from each pair of interacting particles:[1]

$$E = \sum_{<i,j>} g(\mathbf{r}_i, \mathbf{r}_j) \tag{7.6}$$

Here, the function g is typically represented by any kind of empirical or semi-empirical functional form, but in our case it will be discovered by SR. The data for this task comes in the form of snapshots of an atomic system consisting of a list of Cartesian coordinates of its atoms and a value of the total energy.

Let us assume we have an atomic system consisting of n atoms. Since it would not be practical to make use of the atomic coordinates $\mathbf{r}_i$ directly, the raw input coordinates are transformed into a distance matrix $D = [d_{ij}]$, where $d_{ij} = \|\mathbf{r}_i, \mathbf{r}_j\|$ and $i, j = 1, ..., n$:

$$\begin{bmatrix} \mathbf{r}_1 \\ \mathbf{r}_2 \\ \vdots \\ \mathbf{r}_n \end{bmatrix}^{n \times 1} \longrightarrow [d_{ij}]^{n \times n} \tag{7.7}$$

The matrix $[d_{ij}]$ is then used to aggregate pairwise distance around each atom and to compute the total energy according to Equation (7.6).

Considering local energies around each atom as a sum of pairwise interactions has the advantage of preserving the symmetry of the physical system (rotational, translational, and permutational invariances), at the same time leading to models that are not depending on any particular frame of reference.

In order to obtain the energy in the form of Equation (7.6), the typical SR primitive set is extended with a special symbol $\sum$ that can perform

---

[1] Pairwise interactions are also called two-body interactions since two interacting atoms are considered. Depending on application area and modelling goal, more atoms can be considered by the interaction model, however, this comes at the cost of highly increased computational costs.

the summation of atomic distances. The $\sum$ symbol aggregates a selection of pairwise distances into a single value which is then processed further by the model to generate a prediction for the energy. This straightforward approach of summing up pairwise distances was first demonstrated by Hernandez et al. (2019); its advantage is that it allows any kind of functional forms to be evolved by the algorithm without the limitations of a predefined model structure.

The number of interacting neighbors around each atom is determined by inner and outer radii (expressed in Å), therefore the number of neighbors of each atom (and the number of terms in the summation) can vary. The summation symbol Σ is evaluated by taking distances from distance matrix d_{ij} (relative to each atom) and summing them up.

$$\text{sum of all distances} = \sum_{\text{atom } i} \sum_{\text{neighbor } j} d_{ij} \qquad (7.8)$$

Because the Σ symbol performs a sum-reduction from $[d_{ij}]$ to a single scalar value, it is clear that multiple Σ symbols cannot be nested inside the tree structure. To avoid this situation, a simple rule can be implemented in the interpreter used to evaluate expressions: whenever a Σ symbol is found underneath another Σ symbol, the lower symbol acts as the identity function $f(x) = x$. Alternative approaches are for example the use of grammars that forbid nesting of Σ symbols or specialized manipulation operations.

With the custom primitive set including summation symbols we can approach the problem of predicting the total energy of an atomic configuration. The training data used is available online and comes from quantum-chemical calculations of the energy of an atomic cluster consisting of 32 Cu (copper) atoms.[2] From the 150 coordinates–energy snapshots available, half are used for training and half for test. The data is shuffled beforehand.

We employ the NSGA-II algorithm to model this data, using the parameter settings shown in Table 7.9.

The algorithm is able to discover a model that simplifies to the following expression:

$$-\frac{0.207\,\Sigma\big((0.321\,r - 1.272)\,(0.479\,r - 1.101)\,\Sigma\big(0.707 - \frac{1.851}{r}\big)\big)}{\Sigma\big(\big(-0.714 + \frac{1.793}{r}\big)\big(-0.707 + \frac{1.911}{r}\big)\big(-0.482 + \frac{1.443}{r}\big)\big(-0.482 + \frac{1.911}{r}\big)\big)}$$
$$-106.76$$
$$(7.9)$$

where r represents the corresponding distance d_{ij} between atoms i and j.

The model described by Equation (7.9) achieves a $R^2 = 0.973$ on the training set and $R^2 = 0.979$ on the test set.

---

[2]Data freely available in the POET repository at `https://gitlab.com/muellergroup/poet/-/tree/master/poet_run/data` – see also (Hernandez et al., 2019).

Table 7.9: NSGA-II parameters.

Parameter	Value
Population size	1,000 individuals
Max. tree size	20 nodes
Max. tree depth	10 levels
Crossover rate	100%
Mutation rate	25%
Selection	crowded tournament selection (group size 5) (Deb et al., 2002)
Objectives	R^2 and model length
Generations	1,000

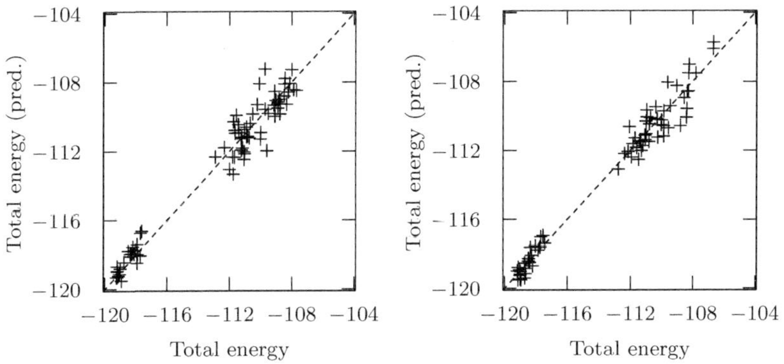

Figure 7.9: Scatter plots for training (left) and test (right) data.

7.4 Friction

In the design of mechanical systems which rely on friction it is important to predict the friction forces which will occur under different loads. To give an example, mechanical brake systems for instance on bicycles, automobiles, or wind turbines must be designed such that it is guaranteed that the braking force is sufficient to stop the moving mass quickly enough under all relevant load scenarios including big moving masses and high velocities. If we overengineer the braking system it is guaranteed to be safe but at the same time it will be unnecessarily heavy, large, and expensive. Therefore, we aim to design the system such that it is safe for all specified loads but not overengineered.

Using simulation models we can predict the behaviour of such mechanical systems accurately especially when friction effects can be neglected. However, if friction plays an important role this becomes more difficult, as friction is a phenomenon that results from forces occurring between the two surfaces sliding against each other which is difficult to capture in a physics-based model. This is partially because many properties such as the roughness of surfaces, strength or hardness of the surface, temperature, wear, loose particles, and properties of lubricants have a large effect and are themselves difficult to quantify. As a consequence, friction models are *phenomenological models* which describe the observed behaviour but are not grounded in a fundamental theory. In this section we will study how we can use SR to build accurate phenomenological models for friction systems.

We study a particular application using data donated by Miba, an Austrian company which develops friction components. The data were collected on a rotational friction testbench in laboratory conditions using a standardized testing procedure. The velocity and pressing force are varied systematically in the experiments and the pressing and pulling forces are measured via calibrated strain gauges. The coefficient of friction for static and dynamic friction is calculated from the measured forces. For each of the load points – a unique combination of velocity and pressure – the test is repeated multiple times, whereby the friction duration is constant for all tests in the experiment. For each repetition, the temperature increases as a consequence of the heat resulting from friction. The temperature is measured and collected in the dataset because it is another variable that affects friction behaviour. The same experiment is repeated for each of the different friction materials.

Figure 7.10 shows the scatter plot matrix for the whole dataset. The plots show that the space of combinations of $p_{\rm rel}$ and $v_{\rm rel}$ is systematically explored. The two targets $\mu_{\rm stat}$ and $\mu_{\rm dyn}$ are highly correlated. T_0 and the temperature increase $T_{\rm inc}$ are also correlated. This correlation occurs because the tests are repeated and a high $T_{\rm inc}$ causes a higher T_0 for the next test. The temperature increase is explained by the mechanical energy that is converted into heat and therefore depends mainly on the friction power $P_{\rm rel} = \mu_{\rm dyn} \, v_{\rm rel} \, p_{\rm rel}$ as the friction duration is constant over all tests. The scatter plots of μ over $p_{\rm rel}$ and $v_{\rm rel}$ do not show a strong correlation. The scatter plots for T_0 and μ show a nonlinear dependency. Increasing temperatures decreases μ.

The Miba dataset has 2016 rows and seven columns. The observations stem from 14 different materials each tested with six pressures, four velocities, and six repetitions at different temperatures ($2016 = 14 \times 6 \times 4 \times 6$). The first column identifies the material (A...N). The three input variables are: $p_{\rm rel}$ a dimensionless pressure value, $v_{\rm rel}$ a dimensionless velocity value, and T_0, the temperature offset from the smallest temperature observed over all experiments in Kelvin. The first tests for each load point have approximately $T_0 = 0$. The two target variables for the prediction are: $\mu_{\rm stat}$, and $\mu_{\rm dyn}$.

The Miba dataset is large enough and allows us to split the dataset 50/50 into training and test partitions. Again we must split carefully because the individual observations are not independent. Therefore, we split the data systematically by load points as shown in Table 7.10. All repeated tests with the same load (p, v) are assigned either to the training or test partition.

There are two options how we handle differences in friction performance

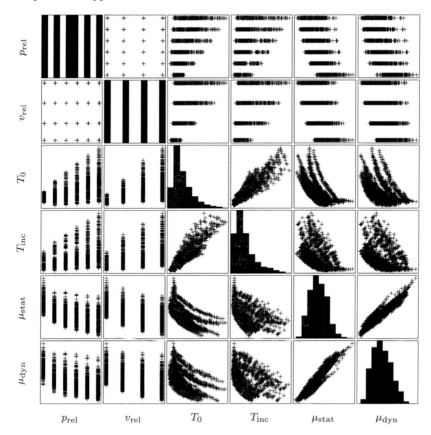

Figure 7.10: Scatter plot matrix for the friction dataset.

for the different materials. Either we build a separate model for each material by fitting it only to a subset of the dataset, or we use SR to find a common model structure with material-specific coefficients. We use the latter and aim to find a common model because we know that the friction behaviour is similar for all tested materials. To accomplish this task we use the material identifier (A...N) as a factor variable (see Section 6.5).

We build two separate models and use p_{rel} and T_0 as input variables for the μ_{stat} model and for the μ_{dyn} model extend the input set with v_{rel}.

The GP parameters listed in Table 7.11 are similar to the hydrodynamics example discussed in Section 7.1. We again vary the complexity of the function set and the maximum size of expressions. The required number of generations is determined from the quality line chart. For the grid search we use fifty generations which should be sufficient to find a good model.

We execute 15 GP runs with different random seeds for each of the configu-

Table 7.10: Assignment of load points to the test partition.

	p					
	0.28	0.57	0.85	1.14	1.42	1.70
$v = 0.26$		Test		Test		Test
$v = 0.71$	Test		Test		Test	
$v = 1.25$		Test		Test		Test
$v = 1.78$	Test		Test		Test	

Table 7.11: GP parameters.

Parameter	Value
Population size	1000
Generations	50
Objective	MSE (minimize)
Memetic optimization	10 Levenberg-Marquardt iterations
Selection	Tournament (group size=5)
Crossover rate	100%
Mutation rate	15%
Replacement	Generational (1 elite)
Max. tree depth	10 levels
Max. tree sizes	$\{10, 15, 20, 25, 30, 35, 40, 45, 50\}$ nodes
Function sets	$F_{\text{Poly}}, F_{\text{RatPoly}}, F_{\text{General}}$

rations and collect the NMSE for the test partition of the training-best model. The 15 runs help to stabilize the estimate of the expected NMSE. Additionally to the function set and the maximum tree size, we also have to select a value for the number of generations. For this purpose, we log the training and test NMSE for the best model on the training set at each generation, and calculate the average test NMSE at each generation over the 15 runs. The generation where we reach the minimum for the average test NMSE is selected as the number of generations for this parameterization. Figure 7.11 visualizes this calculation for two configurations. The minimum average test NMSE is reached after thirty generations for μ_{stat} and already after 17 generations for μ_{dyn}. The plots also highlight that the test NMSE has a high variance which means that our estimate of the expected test NMSE has a high uncertainty.

After we have determined the optimum number of generations for each configuration, we can compare the configurations against each other by analysing the test NMSE values reached at that point. Figure 7.12 and Figure 7.13 show box plots of the test NMSE at the optimal generation number for each configuration. The configuration with the minimum average test NMSE is selected as the best one and used to build the final model using the whole dataset for training. For the static friction model the best configuration is F_{General} with maximally thirty nodes and thirty generations (avg. test NMSE

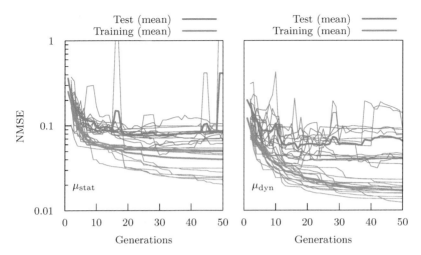

Figure 7.11: Line charts of the NMSE for the training and test set with the best GP configuration (left: μ_{stat}, right: μ_{dyn}). The strong lines mark the average training and test NMSE over generations. The minimum is reached after thirty generations for μ_{stat} and 17 generations for μ_{dyn}.

$= 0.0754$). For the dynamic friction model the best configuration is F_{RatPoly} with maximally 25 nodes and 17 generations (avg. test NMSE $= 0.0531$). The expressions found by GP after simplification are Equation (7.10) for μ_{stat} and Equation (7.11) for μ_{dyn}. The model for the inverse dynamic friction coefficient is easy to interpret because it is almost linear and has only six fitting parameters for each material. The model for the inverse static friction coefficient is more complicated and has many interactions and squared terms when expanded.

$$
\begin{aligned}
\mu_{\text{stat}}^{-1} \approx &\text{Mat}_1 \, (\theta_1 \, p + \theta_2 \, T_0 + \text{Mat}_2) \\
&\left((\text{Mat}_3 \, \log{(\theta_3 \, p)})^{-1} + \text{Mat}_4 \, p + \theta_4 \, T_0 + \text{Mat}_5 \right) + \theta_5
\end{aligned}
\tag{7.10}
$$

$$
\mu_{\text{dyn}}^{-1} \approx (\text{Mat}_1 \, v + \text{Mat}_2 \, p + \text{Mat}_3)^{-1} + \text{Mat}_4 \, p + \text{Mat}_5 \, T_0 + \text{Mat}_6
\tag{7.11}
$$

Figure 7.14 shows the scatter plots of predicted and measured μ_{stat} and μ_{dyn} for the models with training NMSE 2.93% for μ_{stat} and 2.74% for μ_{dyn}.

Figure 7.15a shows the intersection plots for the static friction model over p and T_0. The top row shows the first seven materials, and the bottom row the remaining materials. From the intersection plots it is easy to see that temperature has a much larger effect on static friction than pressure in the identified model. This model predicts that the static friction coefficient decreases with increasing temperature, which conforms to the expected behaviour for the materials tested in this experiment.

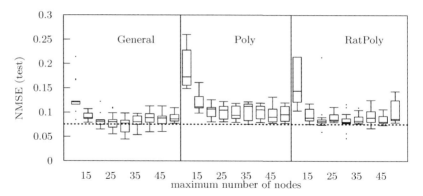

Figure 7.12: Box plots for the test NMSE values for μ_{stat}. The dashed horizontal lines mark the mean of the selected configuration. The configuration with minimum mean test NMSE is F_{General} with thirty nodes.

The intersection plot for the dynamic friction model in Figure 7.15b shows a similar temperature effect. The dynamic friction coefficient decreases with increasing temperature. Additionally, we spot a nonlinear pressure dependency whereby the predicted value decreases with increasing pressure. The effect is stronger for small pressures. For some materials the friction coefficient increases slightly for higher pressures. Finally, changing sliding velocity v only has a small effect.

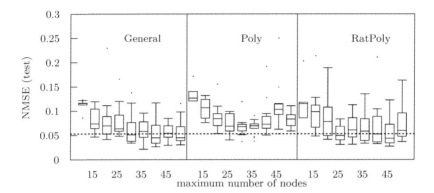

Figure 7.13: Box plots for the test NMSE values for μ_{dyn}. The dashed horizontal lines mark the mean of the selected configuration. The configuration with minimum mean test NMSE is F_{RatPoly} with 25 nodes.

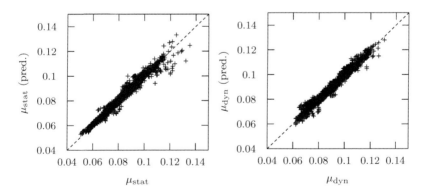

Figure 7.14: Scatter plots for the measured and predicted μ_{stat} and μ_{dyn} for the identified models (Equation (7.10) and Equation (7.11)).

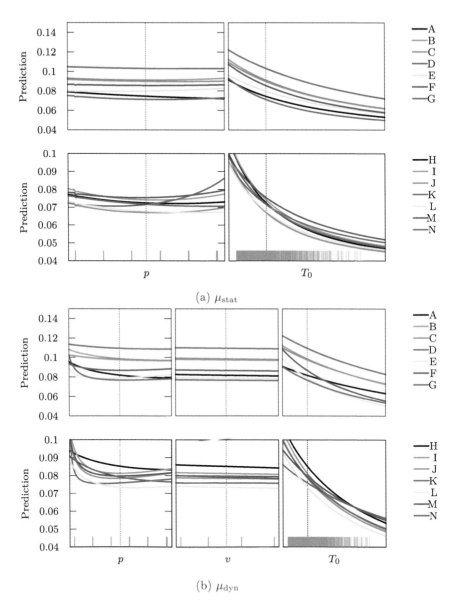

Figure 7.15: Intersection plots for the friction models.

7.5 Lithium-ion Batteries

Today, lithium-ion cells are used to power almost any mobile device and the durability of batteries is one of the most important indicators for the cost-effectiveness of batteries. It can be measured as the number of charge/discharge cycles that can be performed until the battery reaches 80% or 70% of its initial capacity. The capacity of batteries is typically given in Ah (Ampere hours) and can be calculated from the discharge current and duration. Typically the capacity of lithium-ion batteries decreases nonlinearly with the number of charge/discharge cycles. Once a battery reaches 70% or 80% of initial capacity the useful battery lifetime is reached. From this point on, the capacity diminishes rapidly with each cycle.

Another quantity of interest is the state of charge (SoC) which is the remaining capacity or discharge duration of a singe discharge cycle. Effects that influence SoC are mainly the maximum capacity, the discharge current, and the temperature (How et al., 2019) assuming that the battery was loaded to full capacity. An important indicator for SoC is the cell voltage because it decreases continuously while the cell is discharged. Lithium-ion cells have a limit for the minimum voltage and should not be deeply discharged beyond this voltage limit.

7.5.1 NASA PCoE Battery Datasets

Several groups of researchers have run experiments to analyze battery lifetime with the aim to improve battery models. In this example we use the dataset published by the Prognostic Center of Excellence (PCoE) at NASA Ames to demonstrate how SR can be used to find models for lithium-ion batteries.

The dataset contains measurements for 34 lithium-ion cells (type 18650), whereby each cell was tested with different discharge parameters. The charging protocol was the same for all cells. The used test bench allowed to test three cells concurrently and consisted of a controllable charger and controllable discharge load as well as a temperature-controlled chamber for the three cells (Goebel et al., 2008). Several batches of data were published over time. We first use the first batch with cells #5,6,7, and #18 which comes with the following description (Saha and Goebel, 2007):

> "A set of four Li-ion batteries (# 5, 6, 7 and 18) were run through 3 different operational profiles (charge, discharge and impedance) at room temperature. Charging was carried out in a constant current (CC) mode at 1.5A until the battery voltage reached 4.2V and then continued in a constant voltage (CV) mode until the charge current dropped to 20mA. Discharge was carried out at a constant current (CC) level of 2A until the battery voltage fell to 2.7V, 2.5V, 2.2V and 2.5V for batteries 5 6 7 and 18 respectively. Impedance measurement was

carried out through an electrochemical impedance spectroscopy (EIS) frequency sweep from 0.1Hz to 5kHz. Repeated charge and discharge cycles result in accelerated aging of the batteries while impedance measurements provide insight into the internal battery parameters that change as aging progresses. The experiments were stopped when the batteries reached end-of-life (EOL) criteria, which was a 30% fade in rated capacity (from 2Ahr to 1.4Ahr). This dataset can be used for the prediction of both remaining charge (for a given discharge cycle) and remaining useful life (RUL)."

To give an impression of the data, Figure 7.16 shows the measured capacity of the four cells over the number of cycles. The plots show that the capacity of cells, which are discharged to a lower voltage limit deteriorates more quickly, leading to a shorter lifetime. The measurements for cells #5, #6, and #7 were made concurrently starting at the same time. The capacity peaks visible in the plot correlate with larger pauses between consecutive discharge cycles. In these pauses, the cells were able to regenerate, or were charged longer, which leads to a capacity spike.

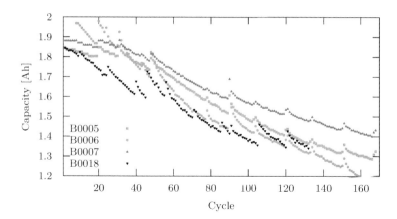

Figure 7.16: Capacity over number of discharge cycles for four cells from the PCoE dataset. The available capacity decreases over cycles. The smaller jumps in capacity are caused by longer pauses within cycles which allow longer charging duration and temporary recovery of cells.

Cell #18 was tested at a later point in time and has peaks at other cycles. This cell was discharged to 2.5 V similar to cell #6. However, the capacity curves for #6 and #18 are very different, and #18 has a shorter lifetime. A closer examination of the data shows that the testing procedure was changed slightly for #18. For the first three cells the time span between discharge cycles is approximately five hours, while for #18 this was reduced to approximately three hours and 45 minutes. The reduced intervals between cycles could be an explanation for the shorter lifetime of #18. Another difference between cell #6 and #18 is a slightly different initial capacity.

All four cells underwent different loads over their lifetime which means that the health state of the cells is not directly comparable even after the same number of cycles. Over the long term however, a linear decrease in capacity is visible for all four cells. Assuming that we are able to measure the actual capacity of the cell via a full charge/discharge cycle, a simple linear model may be sufficient to predict the remaining useful lifetime.

We instead focus on the more difficult task of predicting the state of charge (SoC). Figure 7.17 shows the voltage over time for multiple discharge cycles of cell #5. Initially, the cell lasts approximately 55 minutes with a constant discharge current of 2 A. After 160 cycles the cell lasts less than fourty minutes. This cell reached the end of its lifetime after 168 discharge cycles where its capacity dropped below 1.3 Ah (70%) from initially 1.85 Ah. The voltage curves in discharge cycles shown in Figure 7.17 are nonlinear and monotonically decreasing. Voltage decreases quickly initially, before reaching a plateau, and after some time, the end of the discharge cycle marked by an increasingly faster drop of voltage.

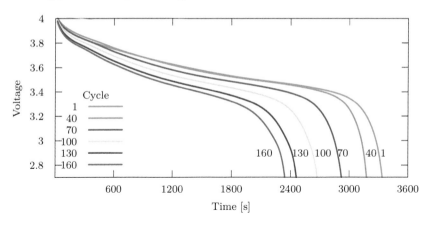

Figure 7.17: Voltage measurements for multiple discharge cycles for cell #5. The voltage curve over discharge duration is highly nonlinear with a sudden voltage drop when the capacity is depleted.

7.5.2 First Version of the State-of-charge Model

We start with a simple model for SoC that shall allow us to predict the remaining discharge duration after ten, twenty, and thirty minutes and assumes a constant current. For this model we only use data from the first batch and use the data for cells #5, 6, 7 for the training phase and data for cell #18 for testing the model. In Section 7.5.3 we will extend the SoC model to support different discharge currents and temperatures which requires us to extend the dataset as well.

In Figure 7.17 we can see that the shape of the voltage curves is similar for all cycles, whereby the discharge duration gets shorter with increasing number of cycles. This indicates that it should be possible to predict the remaining discharge duration based on the voltage measured at a certain time. A lower voltage after for example ten minutes of constant current discharge implies a shorter remaining discharge duration.

We prepared a dataset from the original data with features that might be useful. U is the voltage measured at ten, twenty, and thirty minutes, T is the cell temperature measured at the same time points. n is the number of discharge cycles. C_0 is the initial cell capacity and is calculated as the maximum capacity observed over the first ten discharge cycles for each cell. As the target variable we use the discharge duration until the voltage drops below 2.7 V. Figure 7.18 shows a scatter plot matrix for the dataset with the prepared features. The scatter plot only shows the voltage and temperature after ten minutes as the values for twenty and thirty minutes are similar.

To check the relationship between voltage at a certain time and the remaining discharge duration, we plot the remaining discharge time over the voltage measured after ten, twenty, and thirty minutes for all four cells in Figure 7.19. Except for a few outliers there is a strong nonlinear correlation between the two variables. This implies that it should be possible to predict the remaining discharge duration accurately. All cells except for #6 align well in Figure 7.19. Since the cells #6 and #18 are both discharged to 2.5 V the discharge limit cannot be the cause for the deviation of the data for #6. Closer inspection shows a slightly higher initial capacity for #6 which is a plausible explanation for the deviation and the reason why we included the initial capacity as an additional feature.

For this example, we use a multi-objective algorithm (NSGA-II) to find models with a good tradeoff between goodness-of-fit and size (see Section 4.9.3). Table 7.12 shows the parameter values for the algorithm in detail. When using multi-objective algorithms for SR, we recommend to discretize the fitness component for the goodness-of-fit as this prevents the algorithm from generating many small models with only slightly different goodness-of-fit. We therefore round R^2 to five digits after the comma. More details are given in Section 4.9.3.

The runtime for these settings is only about one minute on a 2022 desktop PC using the software HeuristicLab. We executed only a single run and manually selected a model from the Pareto front. The selected simplified model is shown in Equation (7.12). To improve the numeric stability of the model we shifted U by 3.7 V. The expression is short and has only six coefficients whereby only one of them is nonlinear. In this model, C_0 is the initial capacity, n is the number of discharge cycles and U is the measured voltage. The same structure can be used for the prediction of remaining discharge duration after

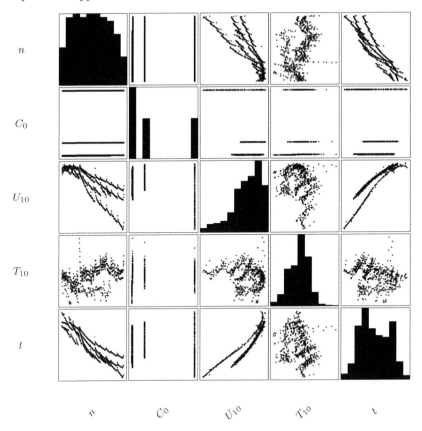

Figure 7.18: Scatter plot matrix for the battery dataset used for the first SoC model.

ten, twenty, and thirty minutes, or any other value between those values, if the coefficients are recalibrated for each case.

The coefficients for the ten minute model are $\theta = (1063, -1.479, 95.05, 19.34, 2889, -396.8)$ and can be interpreted easily to validate the model. The model implies that the remaining discharge duration increases with increased capacity $(\theta_1 > 0)$ and voltage $(\theta_3, \theta_4, \theta_5 > 0)$ and decreases with increasing number of charge/discharge cycles $(\theta_2 < 0)$. These correlations are plausible. The only nonlinear correlation of the prediction is with the voltage.

$$\text{remaining duration} \approx \theta_1 C_0 + \theta_2 n + \theta_3 \exp(\theta_4(U-3.7)) + \theta_5(U-3.7) + \theta_6 \quad (7.12)$$

The intersection plot in Figure 7.20 visualizes the correlations captured by the model and complements the insights that we already gained from

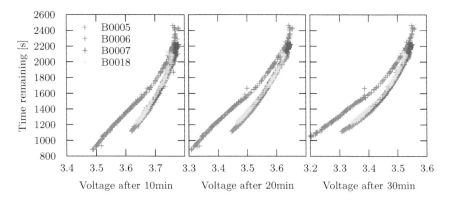

Figure 7.19: Remaining discharge time over voltage after ten, twenty, and thirty minutes of discharge.

interpretation of the coefficients. From the visualization we can see that the correlation between voltage and remaining discharge duration is most dominant.

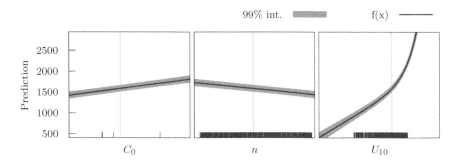

Figure 7.20: Intersection plot for the ten minute model. The model predicts the remaining discharge time in seconds shown on the y-axis.

The predictions produced by the model match the measured values well as shown in the scatter plot in Figure 7.21. The results are similar or even slightly better when the model is recalibrated using the voltage at twenty or thirty minutes as shown in Table 7.13. The model predicts the discharge duration well with a RMSE of thirty to fourty seconds for the training and the test set.

This first SoC model fits the observations well, but is limited to an environment temperature of 24 °C and constant discharge current of 2 A. In the next section we extend our dataset to include more experiments with different temperature conditions and discharge currents and build an extended SoC model.

Table 7.12: NSGA-II parameters for the first SoC model.

Parameter	Value
Population size	500
Generations	100
Initialization	PTC2
Crossover prob.	100% (90% internal nodes)
Mutation prob.	15%
Selection	Crowded tournament with group size 2 (models with equal fitness are dominated)
Maximum tree size	25 nodes
Maximum tree depth	8 levels
Function set	$\{+, -, \times, \div, \exp, \log, x^2, x^3, \sqrt{x}, \text{cbrt}\}$
Terminal set	$\{n, C_0, U, T, \text{coefficients}\}$
Objectives	maximize R^2 and minimize tree size
Parameter optimization	10 Levenberg-Marquardt iterations

Table 7.13: Measures for goodness-of-fit for the simple SoC model (Equation (7.12)).

	Training			Test		
	10min	20min	30min	10min	20min	30min
# parameters	6					
# observations	504			131		
RMSE	29.4	23.8	18.5	43.6	28.3	34.7
R2	0.994	0.996	0.998	0.975	0.990	0.998

7.5.3 Extended Version of the State-of-charge Model

For the extended SoC model, we first prepare a larger dataset by combining data from multiple batches published by the PCoE. In the later experiments, performed from April 2008 until September 2010, the same setup with the same type of cell and the same loading protocol was used, but the discharge current and the environmental temperature were varied.

Table 7.14 shows the cells for which we use data for training and testing the model, together with the parameters controlled in the experiment. We carefully chose the cells to include in our dataset and removed cells which had data issues (e.g., outliers of capacity and voltage or implausible temperature readings). We removed data from the first few discharge cycles because they often contained outliers. Additionally, we manually removed outliers for cycles after longer downtimes and completely excluded cells for which a non-constant discharge current was used.

After the selection of cells to include in our dataset, we systematically assigned them to training and test sets to ensure that both sets include data

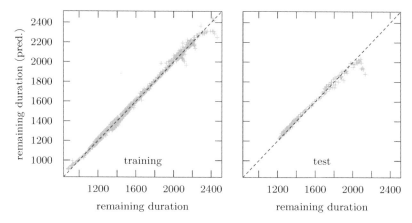

Figure 7.21: Scatter plot of predicted and observed remaining discharge time after ten minutes of discharge for training and test sets.

from experiments with low and high discharge currents as well as low and high temperature. There are three cells with high temperature and discharge current, which is the most critical combination. We use two cells in the training set and one in the test set.

Our approach to modelling will be the same as for the first SoC model. We assume that the cell is fully charged before discharging with a constant current and that we can measure the voltage, current, and temperature. The model shall then be used to predict the total discharge duration based on the measurements up to a certain time in the discharge cycle (ten, twenty, and thirty minutes). Consequently, we prepared the dataset such that one row corresponds to one discharge cycle and each column corresponds to a measurable attribute. The target attribute to be predicted is the total discharge duration t until the discharge limit of 2.7V is reached. As inputs we use the number of discharge cycles n for each cell, the voltage U at the time of prediction after ten, twenty, or thirty minutes of discharge, and the average discharge current I_{avg} and the average temperature T_{avg} for the cycle up to the prediction time. I_{avg} is very close to the nominal discharge current for the cell shown in Table 7.14, T_{avg} is higher because it is the cell temperature which increases on discharge and is in general higher than the environment temperature in the test chamber.

Similar to the first SoC model we can prepare different versions of the dataset to recalibrate the SoC model for the prediction after ten, twenty, or thirty minutes. In the following we only discuss the results for the prediction after twenty minutes. The results are similar for ten minutes or thirty minutes.

Figure 7.22 shows the scatter plot matrix for the attributes in the extended dataset. The bottom row shows the correlation of the target variable t with the inputs, and we already see a strong nonlinear correlation with the voltage at twenty minutes U_{20}. The plot also shows the decreasing discharge duration over the number of cycles n which is caused by the capacity loss of the cell.

Table 7.14: Cells and datasets used for the extended SoC model.

Set	Cell	Temperature [°C]	Current [A]	Discharge limit [V]
Training	#5	24	-2	2.7
Training	#6	24	-2	2.5
Training	#7	24	-2	2.2
Training	#29	43	-4	2.0
Training	#31	43	-4	2.5
Training	#33	24	-4	2.0
Training	#41	4	-1	2.0
Training	#45	4	-1	2.0
Training	#46	4	-1	2.2
Training	#48	4	-1	2.7
Training	#53	4	-2	2.0
Training	#55	4	-2	2.5
Training	#56	4	-2	2.7
Test	#18	24	-2	2.5
Test	#30	43	-4	2.2
Test	#34	24	-4	2.2
Test	#47	4	-1	2.5
Test	#54	4	-2	2.2

The correlation with I_{avg} simply shows that doubling the (negative) current, halves the discharge duration. The negative pairwise correlation of t with T_{avg} is a consequence of the data collection and the correlation between I_{avg} and T_{avg}. For a large (negative) discharge current the cell temperature tends to be higher.

The extended dataset covers a larger range of discharge durations from only about 15 minutes up to ninty minutes, depending on the discharge current. In this scenario, we want to put more weight on model errors for the short cycles and allow larger errors for the longer cycles. In other words we want to minimize the relative absolute error instead of the squared error. An easy way to achieve a similar effect is to fit a model for the log-transformed target variable $\log(t)$ while still using a squared error loss function (as briefly mentioned in Chapter 5). The model for $\log(t)$ can then be back-transformed easily to produce predictions for the original target.

We again use multi-objective GP with NSGA-II with the parameters shown in Table 7.15. In comparison to the parameters used for the first SoC model we now use larger limits for the tree size (50 instead of 25 nodes) and height (12 instead of eight levels) because we expect that we need a slightly larger model to include the additional input variables. We also increase the number of generations (300 instead of only 100) because of the larger search space. The runtime for this configuration with HeuristicLab and a 2022 desktop PC is

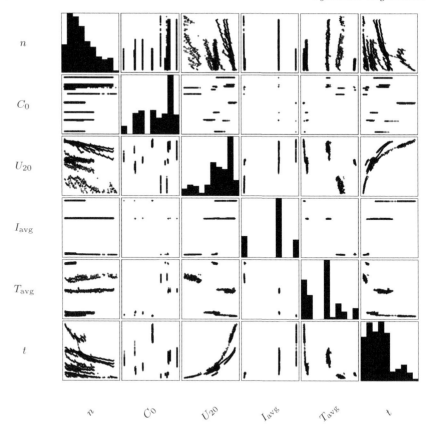

Figure 7.22: Scatter plot matrix for the extended dataset for SoC prediction.

about five minutes. We started several independent runs with different random seeds and selected a single model manually from the Pareto-front with a good tradeoff between goodness-of-fit and size.

The selected model after some algebraic rearrangements to simplify the expression is shown in Equation (7.13).

$$\log(t) \approx \theta_1 T_{\text{avg}} + \theta_2 I_{\text{avg}} + \theta_3 \frac{U_{20}^3 \exp(\theta_4 n)}{\exp(\theta_5 I_{\text{avg}}) + \theta_6 C_0} + \theta_7 \qquad (7.13)$$

The coefficients for the prediction after twenty minutes of discharge are $\theta = (-2.207 \times 10^{-3}, -0.125, 3.71 \times 10^{-2}, -7.263 \times 10^{-4}, -0.2716, -0.2857, 6.44)$. The model has RMSE = 123 seconds for the remaining discharge duration (NMSE = 1.2%, R^2 = 98.8%). Here the importance of the features are harder to interpret because of the more complex rational term which involves almost

Table 7.15: NSGA-II parameters for the extended SoC model.

Parameter	Value
Population size	500
Generations	300
Initialization	PTC2
Crossover prob.	100% (90% internal nodes)
Mutation prob.	15%
Selection	Crowded tournament with group size 2 (models with equal fitness are dominated)
Maximum tree size	50 nodes
Maximum tree depth	12 levels
Function set	$\{+, -, \times, \div, \exp, \log, x^2, x^3, \sqrt{x}, \mathrm{cbrt}\}$
Terminal set	$\{n, C_0, U, I_{\mathrm{avg}}, T_{\mathrm{avg}}, \mathrm{coefficients}\}$
Objectives	maximize R^2, minimize tree size
Parameter optimization	25 Levenberg-Marquardt iterations

all input variables. We only see that temperature has a relatively small linear effect.

The intersection plot for the back-transformed model shown in Figure 7.23 allows us to see the correlations captured by the model more easily. The plot shows that discharge duration decreases with the number of cycles (n) and average cell temperature T_{avg}. Higher capacity also correlates with a longer discharge duration. The strongest correlation is again the voltage, where a higher voltage after twenty minutes means a longer discharge duration. The correlation with the discharge current is also plausible, as the model predicts that the discharge duration is shorter for larger (negative) current. In summary, the correlations captured by the model are all plausible. Another observation is that the prediction interval shown is relatively tight indicating a high confidence of the model.

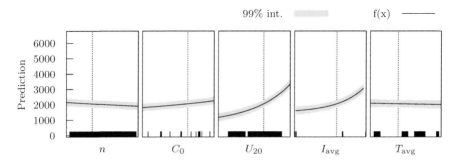

Figure 7.23: Intersection plot for the extended SoC model.

The accuracy of the back-transformed model is also visible in the scatter

plot of actual and predicted durations shown in Figure 7.24. For both the
training and test set, the points are close to the diagonal. Note that the two
datasets are separated such that all measurements from each cell are either
assigned all to the training set or all to the test set. This implies that the
model is able to predict the discharge duration for unseen cells well. However,
the scatter plot shows a slight bias especially for longer discharge durations.
The residuals plot in Figure 7.25 shows that for most of the observations the

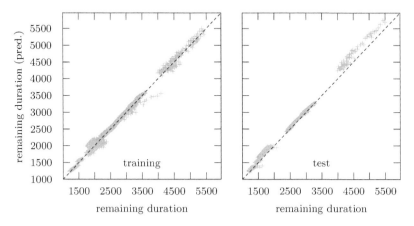

Figure 7.24: Scatter plot of actual and predicted discharge durations for the
extended SoC model for the training and test sets.

prediction is less than 100 seconds off. Some patterns are still visible in the
residuals but their mean is very close to zero almost everywhere which indicates
that there is no systematic uni-variate bias that is not yet captured by the
model.

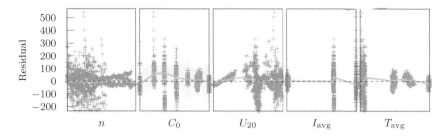

Figure 7.25: Residuals plot for the extended SoC model.

Table 7.16 summarizes the goodness-of-fit on training and test sets. For a
model with only seven fitting parameters it has reaches high accuracy with R^2
values of over 99% on the training set and 98.8% on the test set. The root of
mean squared errors is only 86 seconds on the training set and two minutes on

the test set. This mean error is for discharge durations in the range between 15 and ninty minutes.

Table 7.16: Measures for goodness-of-fit for the extended SoC model shown in Equation (7.13).

	Training	Test
# parameters	7	
# observations	1200	438
RMSE [s]	86.34	122.8
R2	0.992	0.988

In summary, even for the extended dataset captured in experiments with environmental temperatures from 4°C up to 43°C and discharge currents between 1A and 4A, we were able to find a short but highly accurate nonlinear model with only seven fitting parameters using SR. It was not necessary to use a specifically adapted modelling approach or elaborate parameter tuning. A simple multi-objective genetic algorithm (NSGA-II) with rather common parameter settings was sufficient. The runtime was five minutes which is acceptable for this task. No special hardware or powerful GPUs were required as the software runs on a normal desktop PC. The correlations captured by the model are physically plausible and partially nonlinear. The structure of the model includes a nonlinear term which would be hard to construct manually but was automatically identified by SR.

7.5.4 Predicting the Discharge Voltage Curve

For the two SoC models discussed above we used SR without any problem-specific adjustments. With GP it is easy to use problem-specific fitness functions when we require an objective function other than least squares. At its core, GP only requires that the quality of solution candidates can be compared pairwise. Often this is accomplished by assigning a fitness value to each solution candidate and then comparing the fitness values. How the fitness value is determined is completely open to the user and can be implemented for each problem specifically. In some applications, it can even be necessary to run a fine-grained numerical simulation as part of the fitness function.

In this section we use this flexibility to produce another model which allows to predict the voltage curve over the whole duration of discharge. In contrast to the two models described in the previous sections which only allow predicting the remaining discharge duration until reaching the voltage limit, this new model predicts the voltage at each time step of the discharge cycle. We fit this more detailed model to measurements from a single cell only (#5). Once the model structure is known it can be recalibrated for each new cell. This can potentially be done online, starting with an initial guess of model parameters and tuning the parameters after each discharge.

To build this model we require a larger dataset with regular measurements from the whole discharge cycle. So far, we used only aggregated data with a single observation for each discharge cycle. In the new dataset we use the complete data from each discharge cycle with measurements every ten to twenty seconds.

Our goal is to find an expression that we can fit to the voltage data collected up to a certain time and then predict the remaining part of the voltage curve. Figure 7.26 visualizes the idea. In each cycle we collect data from the beginning of the cycle up to the prediction point (e.g., after ten minutes). From this point we use the calibrated model to forecast the voltage curve for the rest of the cycle. Forecasting the voltage is difficult because the model has to correctly forecast the sudden drop in voltage when capacity is depleted based only on the voltage observed in the first ten minutes which is comparatively flat.

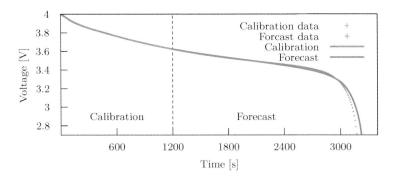

Figure 7.26: The model is calibrated to the first part of the discharge cycle and used to forecast the voltage curve for the second part.

We use a custom fitness function to implement this idea. Each time when a solution candidate is evaluated, we use the expression structure and fit the parameters to the first part of each cycle and predict the remaining part. The parameters are optimized for each individual cycle. The fitness of the solution candidate is the sum of mean squared errors for the calibration data and the forecasts over all cycles. With this procedure we use GP to find an expression structure that best represents the general shape of voltage curves and can be parameterized to match each of the curves. The expression is updated with the optimized parameters for the last cycle to have a good starting point for the next fitness evaluation.

The dataset has three columns: the cycle number n, the discharge duration t, and voltage U and 22,500 rows with observations for 167 cycles. We split the dataset into two partitions, whereby the odd cycles are used for training and the even cycles for testing.

Table 7.17 lists the GP parameter values that we use for this experiment. The procedure requires calibration of model coefficients in each fitness evaluation using nonlinear least squares which is computationally expensive. The runtime for this algorithm was a bit over two days on an 8-core deskop PC. In this time it evaluated about 800,000 solution candidates on a dataset with 11,000 rows. The algorithm stopped after 54 generations because it reached the maximum selection pressure indicating convergence.

Table 7.17: GP algorithm parameters for the voltage curve prediction model.

Parameter	Value
Population size	300
Generations	100
Initialization	PTC2
Crossover prob.	100% (90% internal nodes)
Mutation prob.	15%
Selection	first parent: random
	second parent: proportional
Offspring selection	better than both parents
Maximum selection pressure	100
Maximum tree size	50 nodes
Maximum tree depth	12 levels
Fitness	MSE over calibration and forecast periods
Parameter optimization	50 iterations LM for each cycle
Function set	$\{+, -, \times, \div, \tanh, \exp, \log\}$
Terminal set	$\{U, \text{coefficients}\}$

We use the discharge duration t as the target and U as input variable of the model. As shown in Figure 7.26 the functional dependency between the two variables is monotonic. Therefore, we can use the model in both directions, either to predict t for given U, or to predict U for a given t. Using t as the target variable is more natural, if we want to use the model for SoC prediction. To calculate the predicted discharge duration, we only need to evaluate the model for an input of 2.7 V.

We simply selected the solution candidate with best fitness, i.e., the one with minimal sum of calibration and forecast error on the training set as the solution. The identified expression is shown together with the coefficients for cycle 41 and twenty minutes calibration in Equation (7.14). The result of this experiment was surprising because only three coefficients (β) had different values for each cycle. The remaining six coefficients (θ) were the same over all cycles. This means that GP automatically identified a hierarchical model where six coefficients are used to represent the general shape of the voltage curve and three coefficients are used to calibrate the model to each cycle. This was surprising, because the fitness function has not been designed to encourage this in any way. A possible explanation is that the models, which had only few coefficients optimized for each cycle, had smaller forecast errors overall and were therefore selected more often by GP.

The model has two nonlinear terms which makes it difficult to interpret the coefficients, but the voltage curve for each cycle is a linear combination of the three nonlinear terms.

$$t = \theta_1\,U + \beta_1\,\exp(\tanh(\beta_2\,U\,(\theta_2\,U^2 + \theta_3\,U + 1))) + \beta_3\,\frac{\theta_4\,U + 1}{\theta_5\,U^2 + \theta_6\,U + 1}$$

$$(7.14)$$

$$\boldsymbol{\theta} = (-181.4,\ 0.07103,\ -0.5332,\ -0.2575,\ 0.04084,\ -0.4042)$$
$$\boldsymbol{\beta} = (827.16,\ 50.036,\ 1006.1)$$

Figure 7.27 shows the model output and the measurements for cycles 40, 80, 120, and 160. The plot shows that the model is able to extrapolate well for lower voltage. The voltage drop at the end is predicted well. Table 7.18 lists the parameters and the errors on calibration and forecast periods for the outputs shown in Figure 7.27. The forecast RMSE is between 30 and 45 seconds. The

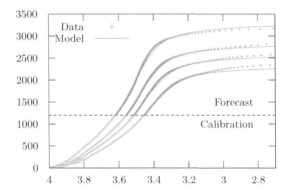

Figure 7.27: Data and prediction of discharge duration in seconds over voltage for test cycles 40, 80, 120, and 160 using Equation (7.14) calibrated to the first twenty minutes for each cycle.

Table 7.18: Parameters and goodness-of-fit for the predictions shown in Figure 7.27.

Cycle	β	Calibration		Forecast	
		R2	RMSE [s]	R2	RMSE [s]
40	(821.6, 50.60, 1005.2)	0.9996	7.0	0.9939	45.0
80	(779.7, 40.85, 771.1)	0.9992	9.5	0.9965	27.6
120	(755.4, 34.37, 7647.2)	0.9988	11.6	0.9930	33.4
160	(731.0, 28.30, 514.3)	0.9986	12.8	0.9908	32.0

linechart in Figure 7.27 only shows predictions for four selected cycles. The scatter plots in Figure 7.28 show the predicted over measured values for all test cycles for the ten minute calibration interval. The RMSE and R^2 values for all calibration durations over all training and test cycles are given in Table 7.19.

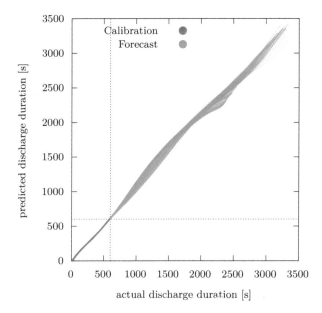

Figure 7.28: Scatter plot of predicted over target discharge durations for all test cycles and the shortest calibration window of ten minutes.

Table 7.19: Medians of goodness-of-fit metrics over all cycles for the voltage curve forecasts (not including the calibration window) for training and test sets.

window size [min]	RMSE (median) [s] Training	Test	R^2 (median) [%] Training	Test
10	6.76	124.36	99.84	96.09
20	9.51	31.92	99.92	99.36
30	11.69	40.74	99.95	97.43
40	18.22	49.72	99.93	92.02

In summary we were able to find short expressions for predicting the state of charge of lithium-ion battery cells with SR. These models allow to predict the remaining duration of a discharge cycle, assuming that the current is constant. An analysis of non-constant current would be an interesting follow up but would require to use a different dataset. The data published by the NASA Prognostic Center of Excellence is mainly for constant discharge currents.

For the first two models we used a multi-objective genetic algorithm to produce short models. The models we presented were taken from the final

Pareto-front and required almost no postprocessing to simplify the expressions. The third model demonstrates the flexibility of the fitness function of GP. Here we used a memetic approach to calibrate a model for each cycle within each fitness evaluation. With GP we evolved the model structure and initial parameters for the calibration. Surprisingly, we found a short hierarchical model. Where only the three parameters were used for calibration in each cycle while the remaining six nonlinear parameters were evolved by GP and held constant for all cycles.

7.6 Biomedical Problems

In this section we present research results achieved in collaboration with the General Hospital (AKH) in Linz, Austria. We have used a medical database consisting of patient data curated at AKH Linz between the years 2005–2008, which includes thousands of samples consisting of 28 routinely measured blood values as well as tumor marker values (varying on a patient basis). In total, anonymized information about 20,819 patients is stored in 48,580 samples.

A tumor marker is a biomarker whose concentration in blood, urine, or body tissue may be suggestive of the presence of certain kinds of cancer. There are several different tumor markers that are used in oncology to this purpose, although by themselves elevated tumor marker values are not sufficient for a diagnosis.

The main modelling tasks addressed here are illustrated in Figure 7.29. Tumor markers are modeled using standard blood parameters and tumor marker data; tumor diagnosis models are trained using standard blood values, tumor marker data, and diagnosis information, and alternatively we also train diagnosis estimation models only using standard blood parameters and diagnosis information.

We consider the following tumor markers: *AFP*, *CA 125*, *CA 15-3*, *CEA*, *CYFRA*, and *PSA* which are summarized below:

- *AFP*: alpha-fetoprotein (Mizejewski, 2001) is a protein found in the blood plasma; during fetal life it is produced by the yolk sac and the liver. In humans, maximum AFP levels are seen at birth; after birth, AFP levels decrease gradually until adult levels are reached after 8 to 12 months. Adult AFP levels are detectable, but usually rather low. For example, AFP values of pregnant women can be used in screening tests for developmental abnormalities as increased values might for example indicate open neural tube defects, decreased values might indicate Down syndrome. AFP is also often measured and used as a marker for a set of tumors, especially endo-dermal sinus tumors (yolk sac carcinoma), neuroblastoma, hepatocellular carcinoma, and germ cell tumors (Duffy et al., 2008). In general, the level

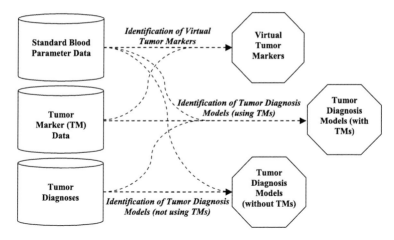

Figure 7.29: Modeling tasks investigated in this research work: tumor markers are modeled, and tumor diagnosis models are trained using standard blood values, diagnosis information, and optionally tumor marker data.

of AFP measured in patients often correlates with the size/volume of the tumor.

- *CA 125*: cancer antigen 125 (CA 125) (Yin et al., 2002), also called carbohydrate antigen 125 or mucin 16 (MUC16), is a protein that is often used as a tumor marker that may be elevated in the presence of specific types of cancers, especially recurring ovarian cancer (Osman et al., 2008). Still, its use in the detection of ovarian cancer is controversial, mainly because its sensitivity is rather low (as documented by Rosen et al. (2005)). Only 79% of all ovarian cancers are positive for CA 125) and it is not possible to detect early stages of cancer using CA 125. Even though CA 125 is best known as a marker for ovarian cancer, it may also be elevated in the presence of other types of cancers; for example, increased values are seen in the context of cancer in fallopian tubes, lungs, the endometrium, breast, and gastrointestinal tract.

- *CA 15-3* mucin 1 (MUC1), also known as cancer antigen 15-3 (CA 15-3), is a protein found in humans; it is used as a tumor marker in the context of monitoring certain cancers (Niv, 2008), especially breast cancer. Elevated values of CA 15-3 have been reported in the context of an increased chance of early recurrence in breast cancer (Keshaviah et al., 2007).

- *CEA*: carcinoembryonic antigen (CEA; (Gold and Freedman, 1965)) is a protein that is in humans normally produced during fetal development. As the production of CEA usually is stopped before birth, it is usually not present in the blood of healthy adults. Elevated levels are seen in the blood or tissue of heavy smokers; persons with pancreatic carcinoma, colorectal carcinoma, lung carcinoma, gastric carcinoma, or breast carcinoma, often have elevated CEA levels. When used as a tumor marker, CEA is mainly used to identify recurrences of cancer after surgical resections.

- *CYFRA*: fragments of cytokeratin 19, a protein found in the cytoskeleton, are found in many places of the human body; especially in the lung. In malign lung tumors high concentrations of these fragments, which are also called CYFRA 21-1, are found. Due to elevated values in the presence of lung cancer, CYFRA is often used for detecting and monitoring malign lung tumors. Elevated CYFRA values have already been reported for several different kinds of tumors, especially for example in stomach, colon, breast, and ovaries. The use of CYFRA 21-1 as a tumor marker has for example been discussed by Lai et al. (1999).

- *PSA*: prostate-specific antigen (PSA) (Andriole et al., 2009; Thompson et al., 2004)) is a protein produced in the prostate gland; PSA blood tests are widely considered the most effective test currently available for the early detection of prostate cancer since PSA is often elevated in the presence of prostate cancer and in other prostate disorders. Still, the effectiveness of these tests has also been considered questionable since PSA is prone to both false positive and false negative indications: according to Thompson et al. (2004), 70 out of 100 men with elevated PSA values do not have prostate cancer, and 25 out of 100 men suffering from prostate cancer do not have significantly elevated PSA.

The documented tumor marker values are classified as "normal" (class 0), "slightly elevated" (class 1), "highly elevated" (class 2), and "beyond plausible" (class 3); this classification is based on medical knowledge. In principle, our goal is to design classifiers for assigning samples to one of these classes. However, here we have decided to produce classifiers that classify samples as "normal (belonging to class 0)" or "elevated (belonging to class 1, 2, or 3)", which simplifies the task to a binary classification problem.

We employ SR with offspring selection as described in Section 4.9.1. The data is split into a training and test partition, and a part of the training data is reserved for model validation. The algorithm returns those models that perform best on validation data. The modelling results summarized in this section have been described previously by Winkler et al. (2010).

7.6.1   Identification of Virtual Tumor Markers

Before using the data in the database to train classifiers for tumor markers, several preprocessing steps are applied to ensure consistency:

1. *Scaling and clipping*: a maximum plausible value $\max_{\text{plau}}(v_i)$ is determined for each input variable v_i and used to linearly scale the values to the unit interval. All values greater than the maximum plausible value are replaced with 1.0.

2. *Merging*: all samples belonging to the same patient with not more than one day difference with respect to the measurement data are merged into a single sample. This step helps to decrease the number of missing values in the data matrix. In rare cases, more than one value might thus be available for a certain variable; in such a case, the first value is used.

3. *Grouping*: to prevent patient data from being included in the training as well as the test data, all measurements are rearranged and clustered sample-wise according to patient ID.

In the next step, separate datasets are compiled for each target tumor marker tm_i by performing the following steps in sequential order:

1. Retrieve all samples containing values for tm_i into a new dataset.

2. Remove features with more than 20% percentage of missing values.

3. Remove any remaining samples that still contain missing values.

This procedure results in a specialized dataset D_{tm_i} for each tumor marker tm_i. An overview of the datasets is given in Table 7.20.

We employ five-fold cross-validation and evaluate the learned classifiers on the entire data at the end of the training process. The results are given in Table 7.21 as the average classification accuracy and their standard deviations on training as well as test data.

Table 7.20: Overview of the datasets compiled for selected tumor markers.

Marker Name (Total Obs.)	Input Variables	Distribution of Samples			
		Class 0	Class 1	Class 2	Class 3
AFP (2 755)	AGE, SEX, ALT, AST, BUN, CH37, GT37, HB, HKT, KREA, LD37, MCV, PLT, RBC, TBIL, WBC	2 146 77.9%	454 16.5%	64 2.3%	91 3.3%
CA 125 (1 053)	AGE, SEX, ALT, AST, BUN, CRP, GT37, HB, HKT, HS, KREA, LD37,MCV, PLT, RBC, TBIL, WBC	532 50.5%	143 13.6%	84 8.0%	294 27.9%
CA 15-3 (4 918)	AGE, SEX, ALT, AST, BUN, CBAA,CEOA, CLYA, CMOA, CNEA, CRP, GT37, HB, HKT, HS, KREA, LD37, MCV, PLT, RBC, TBIL, WBC	3 159 64.2%	1 011 20.6%	353 7.2%	395 8.0%
CEA (5 567)	AGE, SEX, ALT, AST, BUN, CBAA, CEOA, CLYA, CMOA, CNEA, CRP, GT37, HB, HKT, HS, KREA, LD37, MCV, PLT, RBC, TBIL, WBC	3 133 56.3%	1 443 25.9%	492 8.8%	499 9.0%
CYFRA (419)	AGE, SEX, ALT, AST, BUN, CH37, CHOL, CRP, CYFS, GT37, HB, HKT, HS, KREA, MCV, PLT, RBC, TBIL, WBC	296 70.6%	37 8.8%	36 8.6%	50 11.9%
PSA (2 366)	AGE, SEX, ALT, AST, BUN, CBAA, CEOA, CHOL,CLYA, CMOA, CNEA, CRP, GT37, HB, HKT, HS, KREA, LD37, MCV, PLT, RBC, TBIL, WBC	1 145 48.4%	779 32.9%	249 10.5%	193 8.2%

Table 7.21: Accuracy in percent of models identified for selected tumor markers.

Tumor Marker	Accuracy (training)	Accuracy (test)
AFP	77.1 ± 1.4	77.3 ± 2.8
CA 125	71.9 ± 1.7	68.1 ± 8.1
CA 15-3	73.5 ± 1.2	71.7 ± 3.6
CEA	67.9 ± 3.4	65.3 ± 0.7
CYFRA	75.7 ± 3.3	67.6 ± 6.9
PSA	68.6 ± 9.0	65.3 ± 7.7

7.7 Function Approximation

In Chapter 3 we have already seen that SR can be used to find approximations for special functions for which a closed-form algebraic expression does not exist such as the Gamma function

In the following we give a few more examples for the approximation or simplification of expressions. The following examples are taken from Stoutemyer (2013) where some results achieved with the SR software Eureqa are reported and it is discussed how SR tools and computer algebra systems (CAS) could support each other. The first example given by Stoutemyer (2013) is

$$
\cos(x)^3 \sin x + \frac{\cos(x)^3 \sin x}{2} + 2 \cos(x)^3 \cos(2x) \sin x +
$$
$$
\frac{\cos(x)^3 \cos(4x) \sin x}{2} - \frac{3}{2} \cos x \sin(x)^3 - 2 \cos x \cos(2x) \sin(x)^3
$$
$$
- \frac{\cos x \cos(4x) \sin(x)^3}{2} \quad (7.15)
$$

which can be simplified to

$$
2 (\sin(3x) - \sin x) \cos(x)^5 \quad (7.16)
$$

With SR software it is possible to find an even shorter expression. It takes only a few seconds to find Equation (7.17) with effectively zero error on the training set with Operon. This is the same expression as the one found with Eureqa as reported in Stoutemyer (2013).

$$
f(x) = \sin(4x) \cos(x)^4 \quad (7.17)
$$

To find this solution we calculate the result of Equation (7.15) for 129 regularly spaced x values in the range $[-\pi \ldots \pi]$. For this it is recommended to use a software library to calculate the function result with high precision floating point operations. We use the resulting dataset as input for SR. The GP parameters for this experiment are summarized in Table 7.22.

Another interesting example found in Stoutemyer (2013) is Equation (7.18) which is a function defined via an integral.

$$
W(x) = \int_0^x \frac{dt}{\sqrt{(1-t)^2 + \frac{2}{\pi} \cos(\pi \frac{t}{2})}} \quad (7.18)
$$

SR enables us to find an approximation for this expression in closed-form. To prepare a dataset for this task, we use a high-quality numeric solver for the differential equation to calculate the result for each x in the interval $[-1, \ldots, 1]$ with step size $\frac{1}{64}$. We use a similar parameterization as shown in Table 7.23

Table 7.22: GP parameters used for SR for simplification of Equation (7.15).

Parameter	Value
Population size	10,000
Maximum generations	100
Maximum size	10 nodes
Mutation prob.	25%
Crossover prob.	100%
Selection	Tournament (group size 7)
Parameter optimization	10 Levenberg-Marquardt iterations
Fitness	negative MSE
Function set	$F = \{+, \times, \div, \log, \exp, \sin, \cos, \tan, x^2, x^{1/2}\}$
Terminal set	$T = \{\text{coefficient} \times x, \text{coefficients}\}$

Table 7.23: GP parameters used for SR for simplification of Equation (7.18).

Parameter	Value
Population size	10,000
Maximum generations	100
Maximum size	20 nodes
Mutation prob.	25%
Crossover prob.	100%
Selection	Tournament (group size 7)
Parameter optimization	10 Levenberg-Marquardt iterations
Fitness	negative MSE
Function set	$F = \{+, \times, \div, \log, \exp, \sin, \cos, \tan, \text{asin}, \text{acos}, \text{atan}, \sinh, \cosh, \tanh, x^2, x^{1/2}, x^{1/3}\}$
Terminal set	$T = \{\text{coefficient} \times x, \text{coefficient}\}$

but extend the function set to include more trigonometric functions that might be useful to express this function.

The expression found by GP after algebraic simplification is shown in Equation (7.19).

$$W(x) \approx 0.8255 \left(\arcsin x - 0.5161\, x\right) + \tan\left(0.1957\, x\right) + \tan\left(0.1867\, x\right) \quad (7.19)$$

This expression has an average absolute relative error of 0.04% and a maximal relative error of 0.19% for the full input range $x \in [-1\ldots 1]$. The approximation is far from perfect, but it can still be useful to have a closed-form expression which uses only a set of standard functions that are available on most computer systems.

Figure 7.30 shows the original function as well as the approximation in the left panel and the absolute relative error of Equation (7.19) in the right panel.

SR can also be used to find approximate solutions to differential equations

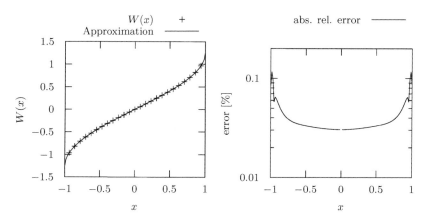

Figure 7.30: The function $W(x)$ and Equation (7.19) found via SR and the abolute relative error of of the approximation (right).

as already shown by Koza (1992). To demonstrate this capability we take an example from Zill and Cullen (2005) shown in Equation (7.20).

$$\frac{dy}{dt} = -3\,y + 13\,\sin(2\,t) \tag{7.20}$$

We use a numeric solver to generate a training set using $y(0) = 6$ and 101 points for $t \in [0, 10]$ with steps of $\frac{1}{10}$. Again we use a high-precision numeric solver and very small tolerances (absolute and relative tolerance $< 10^{-12}$) to make sure that we have accurate training data which is required to find accurate approximations. The GP parameters are shown in Table 7.24. Again we only allow a small number of nodes to enforce a short expression. We use a rather generic function set and include trigonometric functions because the differential equation already contains $\sin(x)$. Only a few generations are necessary to find the expression in Equation (7.21) with RMSE $< 10^{-20}$.

$$\begin{aligned} \hat{y}(t) =&\, 3.605 \left(-\sin\left(1.999\,t - 3.729\right) + e^{0.7969 - 2.999\,t} \right. \\ &\left. - 8.328 \times 10^{-13}\,t + 1.297 \times 10^{-11} \right) \end{aligned} \tag{7.21}$$

This expression can be expanded using a computer algebra system to Equation (7.22) and after some algebra and rounding we get the solution in Equation (7.23) which corresponds to the optimal analytical solution.

$$\hat{y}(t) = 3.605 \left(0.832\,\sin\left(1.999\,t\right) - 0.5547\,\cos\left(1.999\,t\right) + e^{0.7969 - 2.999\,t} \right) \tag{7.22}$$

$$\hat{y}(t) = 3\,\sin\left(2\,t\right) - 2\,\cos\left(2\,t\right) + 8\,e^{-3\,t} \tag{7.23}$$

Figure 7.31 compares the numerical solution and the SR model. In the left

Table 7.24: GP parameters used for finding a closed-form solution for Equation (7.20).

Parameter	Value
Population size	1000
Max. generations	25
Max. size	15 nodes
Mutation prob.	25%
Crossover prob.	100%
Selection	Tournament (group size 5)
Parameter optimization	10 Levenberg-Marquart iterations
Fitness	negative MSE
Function set	$F = \{+, \times, \div, \log, \exp, \sin, \cos, x^2, x^{1/2}\}$
Terminal set	$T = \{\text{coefficient} \times x, \text{coefficient}\}$

panel the data points taken from the numerical solution are shown together with the SR solution. In the right panel the relative absolute error of the SR solution (Equation (7.21)) and the true solution (Equation (7.23)) are plotted. Even the true solution deviates slightly from the numerical solution because of inaccuracies of the solver and of floating point calculations.

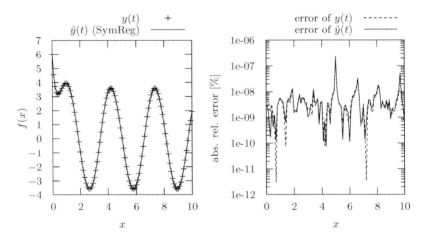

Figure 7.31: The numerical solution for $y(t)$ and the solution for the SR model (Equation (7.22)) with $y(0) = 6$. The plot on the right shows the absolute relative errors of the numerical solution (Equation (7.20)) to the SR solution (Equation (7.22)) and the true solution (Equation (7.23)).

7.8 Atmospheric CO_2 Concentration

The concentration of CO_2 in the atmosphere is an important factor for global warming and has been increasing steadily with the industrialization of global economy. The longest record of direct measurements of CO_2 in the atmosphere has been collected at the Mauna Loa Observatory in Hawaii and is available online.[3] The dataset contains monthly averages of CO_2 in mole fractions (micromole/mole, ppm) starting with March 1958 and exhibits a long term increasing trend as well as a yearly periodic pattern as shown in Figure 7.33. The time series of measurements has a few missing values which have been interpolated by the data provider. For this example we have used these interpolated values.

Using SR and GP we can identify an autoregressive model to generate a prediction for the atmospheric CO_2 concentration for the next 15 years. Our autoregressive model for the CO_2 concentration has the form

$$y_{t+1} = f(y_t, \ldots y_{t-k}, \text{decimal\_date}_t, \boldsymbol{\theta}) + \epsilon_t$$

where y_t is the observed value of the time series at time t, k is the maximum lag, decimal_ date is a floating point value that encodes the year in the integer part and the month in the fractional part, and $\boldsymbol{\theta}$ is the vector of fitting parameters. In this example the we have monthly observations and $k = 23$ means that we may use the observed value of the current month y_t and the previous 23 months to predict the value for the next month y_{t+1}. After fitting the parameters to the time series, the model can be used for forecasting by feeding back predicted values as input values recursively (see Section 2.4). The recursive procedure starts with the observed values for $y_t, \ldots, y_{t-k}$ for the prediction of y_{t+1}. Then all values are shifted by one time step $t \leftarrow t + 1$, and the predicted value becomes the new current value for the prediction of y_{t+2}. In this procedure any error made by the model is amplified by the repeated application of the model. Thus, the prediction interval becomes wider for predictions that are further in the future.

For our experiment with GP we use the function set $\{+, -, \times, \div, \log, \exp\}$ and $k = 23$, which means that the terminal set contains $y_t, \ldots, y_{t-23}$ as well as coefficients. Again, we do not fix the model structure, or the lagged values that must occur in the model. Instead, we rely on GP to identify the model structure and to select the necessary lagged values automatically. The maximum size of the model is set to 25 tree nodes. For training the model we use the observations from 1962 to 2008 and minimize the sum of squared errors.

After a few seconds, GP found the expression tree shown in Figure 7.32.

---

[3]Dr. Pieter Tans, NOAA/GML (gml.noaa.gov/ccgg/trends/) and Dr. Ralph Keeling, Scripps Institution of Oceanography (scrippsco2.ucsd.edu/).

This expression tree can be algebraically simplified to

$$y_{t+1} \approx \theta_1\, y_{t-10} + \theta_2\, \text{decimal\_date}_t + \theta_3\, y_t + \theta_4\, y_{t-1} + \theta_5 \qquad (7.24)$$
$$\boldsymbol{\theta} = (0.39766,\ -0.0010935,\ 0.89874,\ -0.28848,\ -0.014739)$$

This linear autoregressive model is made up of the three lagged terms y_{t-10}, y_{t-1}, y_t as well as the date. This model has a one-step ahead prediction

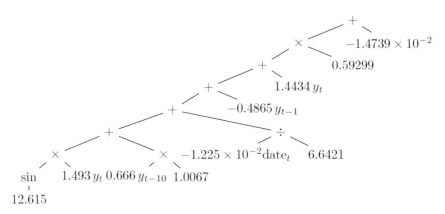

Figure 7.32: The expression tree evolved by GP for the CO_2 dataset. The expression can be algebraically simplified to the linear autoregressive model (Equation (7.24)).

error of 0.093% (RMSE=0.40ppm) on the training set (1962 – 2008) and 0.087% (RMSE=0.44ppm) for the test set (2008 – 2020). Figure 7.33 shows the measurements and predictions of the model for the time frame used for testing the model (2008 – 2020).

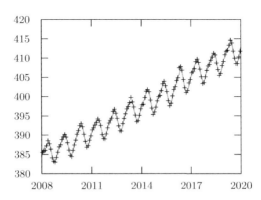

Figure 7.33: Measurements and predictions of the CO_2 model for the test set.

In Section 6.1.1 we already discussed a simple static model for the CO_2

concentration and described how prediction intervals can be calculated for SR models. For the autoregressive model, we cannot use the simple linear approximation discussed in Section 6.3, because we have to account for the uncertainty of the predictions which are fed back into the model as inputs. Instead, we can use a Monte Carlo technique to estimate the uncertainty. Figure 7.34 shows multiple predictions for the test period (2008 – 2020) as well as a forecast until 2035 sampled from the fitted model (Equation (7.24)). To visualize the uncertainty of the model, we used a Laplace approximation for the posterior distribution for the coefficient vector around the maximum likelihood estimate and sampled hundred coefficient vectors randomly from this distribution. For each of the sampled coefficient vectors we produced the forecast of the autoregressive model starting from the same initial value. Each line in the plot is one such forecast. In a similar way it would be possible to account for the uncertainty from multiple models by Bayesian averaging.

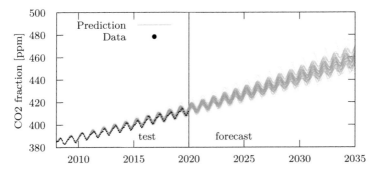

Figure 7.34: Predictions of the atmospheric CO_2 model for 2008 to 2035 using Monte Carlo simulation an coefficient vectors sampled from the Laplace approximation for the posterior distribution of the model coefficients.

As shown in Figure 7.34 the autoregressive model identified by GP predicts the observations in the test set well and the uncertainty is small. When predicting further into the future, the trend of is continued and the uncertainty increases as expected. The seasonal component of the observations is captured well even for the long-term forecast.

7.9 Flow Stress

The stress-strain curve describes the force required in deformation of materials. Different materials exhibit different stress-strain curves, whereby the atomic structure of the material is the most important factor for the deformation behaviour. Additionally, processing conditions such as the deformation rate and temperature also affect the deformation process. Understanding the stress-strain behaviour is for instance important in all kinds of metal forming processes including forging, bending, or extrusion. Predicting the stress-strain curve for a given metallic material with known chemical composition can also be helpful for the development of new metallic alloys with specific properties.

In the following, we give an example of using SR for creating a model of the stress-strain curve for a well-known aluminium alloy (AA6082), which allows to predict the stress-strain curve for deformation at different temperatures. We use a dataset collected from multiple hot compression tests with controlled temperatures in the range from 200 to 500 °C in steps of 25°C. Multiple deformation rates were used in the full experiment (Kabliman et al., 2021), but here we use only the data for a single deformation rate ($\dot{\epsilon} = 0.1$). This dataset contains twelve curves (one for each hot compression test at a controlled temperature) and is shown in Figure 7.35. For each curve we use exactly 600 data points for ϵ in the range 0% to 60% with increments of 0.1%. Each curve shows the required force in the test over the deformation ratio ϵ (true strain).

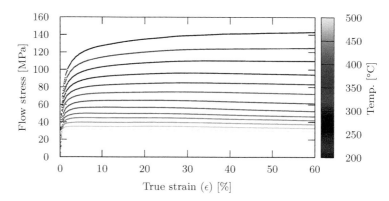

Figure 7.35: Stress-strain curves for aluminium alloy 6082 at deformation rate $\dot{\epsilon} = 0.1$ at different temperatures.

The phase of plastic deformation occurs at the beginning of the deformation process. In this phase the flow stress increases linearly, which is hard to see in Figure 7.35, because the behaviour quickly shifts to non-plastic deformation for this material. This shift occurs at a certain deformation ratio ϵ and is temperature dependent. Even though there are physics-based models, they often have calibration parameters that have to be fitted to measurements, and the models are limited to only a few well-understood alloys. SR could potentially be helpful to build a statistical model for stress-strain solely from data which allows to produce accurate stress-strain models for new alloys quickly. All the techniques for knowledge integration discussed in Section 6.1 can be used to improve the efficiency of the process or physical validity of the models. Additionally, SR provides the potential to produce interpretable models which can be inspected and analyzed easily. The formula can also be used easily within commercial software tools for the simulation of deformation processes.

Because the measurements of stress-strain form a time series, subsequent observations are not independent and we have to partition the dataset carefully. Thus, we order the data by temperature and use the complete data of every second hot compression test for training. The resulting training set contains data from six tests. The data from the remaining seven tests are used for testing the predictive performance of the finally selected model. By using every other curve for testing the model, we get a better estimate of the model prediction error for new temperatures.

The straightforward way to build a model for this dataset is to use the approach known from earlier chapters of this book to find a static regression model for flow stress kf

$$\text{kf} \approx f(T, \epsilon, \boldsymbol{\theta})$$

with kf measured in MPa as a function of temperature T, deformation ratio ϵ, and a vector of coefficients $\boldsymbol{\theta}$. However, since the data captured in the hot compression test is effectively a time series, where the force and the deformation is measured at each time step, we instead try to identify a model in the form of an ordinary differential equation

$$\frac{\partial \hat{\text{kf}}}{\partial \epsilon} = f(\hat{\text{kf}}, \epsilon, T, \boldsymbol{\theta})$$

to simulate the dynamics of the deformation process. Solving the ODE numerically with assumed initial value $\hat{\text{kf}}(0) = 0$ provides $\hat{\text{kf}}(\epsilon)$, which we can evaluate for all ϵ_i in the training set and we minimize the sum of SSE

$$\sum_T \|\hat{\text{kf}}_T(\epsilon_i) - \text{kf}_{T,i}\|_2^2 \tag{7.25}$$

between the simulated $\hat{\text{kf}}_T(\epsilon_i)$ and the measurements $\text{kf}_{T,i}$ over all tests, whereby each test is for a different temperature T. We have already described how we can use SR to find differential equations and how to optimize the coefficients $\boldsymbol{\theta}$ in Section 6.4. In contrast to the static model, the dynamic model allows us to simulate a stress-strain curve with time-varying temperature. Additionally, as we have mentioned earlier, it is often more natural to model dynamic processes using differential equations, which can lead to more compact and interpretable models.

Again, we use tree-based GP with NSGA-II to minimize tree size and sum of squared errors with population size = 1000 for 100 generations. In every generation, 2000 parents are selected with crowded tournament selection with group size two, and 1000 new solution candidates are generated using single-point crossover with a probability of 90% and subsequent mutation with a probability of 15%. Within fitness evaluation, coefficients are optimized using 100 L-BFGS iterations and 10 random restarts. After optimization, the coefficients of solution candidates are updated with the coefficient vector providing the best sum of squared errors. The 1000 best solution candidates from the union of the parent generation and the newly generated solution candidates build the population for the next generation.

The runtime for this configuration is approximately one day on an office computer, whereby most of the total runtime is spent in optimization of the coefficients, which requires numerically solving an ODE for each optimization step. In total approximately 100,000 equations are generated. With 10 random restarts, and a maximum of 100 iterations for optimization of coefficients, this requires 100 million calls to the ODE solver.

One of the equations from the Pareto-front of the final generation is

$$\frac{\partial \hat{\text{kf}}}{\partial \epsilon} = \frac{\theta_1 \, \hat{\text{kf}}}{\theta_2 \, T + \theta_3 \, \hat{\text{kf}} + 1000 \, \epsilon + \theta_4} + \theta_5 \exp(\theta_6 \, \hat{\text{kf}}) + \theta_7 \, \epsilon + \theta_8 \, \hat{\text{kf}} + \theta_9 \, T + \theta_{10}$$

(7.26)

$$\boldsymbol{\theta} = (-3.147, -0.2754, -1.278, 299.5, 15.49, -0.02228,$$
$$- 1.681 \times 10^{-3}, -0.03381, -0.04137, 14.06)$$

This relatively simply model has RMSE = 1.17 MPa on the training set and RMSE = 1.58 MPa on the test set. Figure 7.36 shows the measured and predicted stress-strain curves for the training and test sets. The model fits the data well even for the test set except for the highest temperatures where the model prediction is biased low.

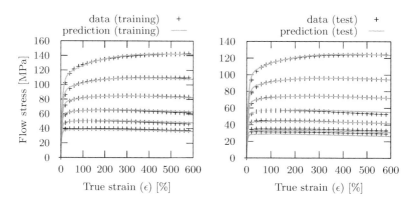

Figure 7.36: Measured and predicted stess-strain curves for training and test sets.

With this result we conclude this example and will not go into further details at this point. More details about modelling stress-strain curves with SR are given for instance by Versino et al. (2017). The dataset discussed in this section is described in more detail by Kabliman et al. (2021) including SR results for the full dataset. Using SR for the development of data-driven plasticity models is also discussed in more detail by Bomarito et al. (2021).

7.10 Dynamics of Simple Mechanical Systems

In Chapter 3 we have already shown that it is possible to find simple physical laws from experimental data. In this section we revisit this topic, and show how SR can be used to find models in form of explicit ordinary differential equations (ODE) for simple mechanical systems. We discuss four examples of increasing complexity: the linear oscillator, the linear double oscillator, the pendulum, and the double pendulum. The examples and datasets are taken from Schmidt and Lipson (2009).

These datasets contain measurements acquired via a motion tracking system from the physical systems in a laboratory setting. Since data are acquired from the physical systems, the data are potentially noisy and damping effects through friction can occur. Details on data acquisition and preprocessing can be found in the article by Schmidt and Lipson (2009). Additionally, to the "real-world" datasets synthetic datasets from numeric simulation of these systems are also available. In the following, we use the synthetic datasets for the simulated double pendulum and measurements from the physical systems for the remaining systems. Our goal in all four cases is to find a system of ODEs using only the measurements and numeric approximations for the derivatives of state variables which are already contained in the datasets. We do not use additional preprocessing steps such as smoothing the approximate derivatives.

Our approach is different from the approach discussed by Schmidt and Lipson (2009) where the goal is the identification of implicit form differential equations, as we focus on the identification of equations of motion in ODE form.

The approach for identifying the models is the same for all four systems. The first two-thirds of the dataset are used for training and the last third is used as a test set. The target variables are the (approximated) derivatives of the state variables. All state variables are allowed as inputs and fitness is the MSE. Table 7.25 shows the GP parameters we use for all experiments, except for the double pendulum, where we used a configuration with larger settings. We first determine the necessary number of nodes, the function set, and the number of generations for each system with a grid search and multiple random restarts. The configuration with the smallest number of nodes and an average test MSE within one standard deviation of the average test MSE of the best configuration is selected and used to produce the final model.

For the double oscillator and the double pendulum we identify the two equations independently and only combine them at the end for simulation. In the end we solve the system of ODEs numerically for the initial values from the datasets for the training and test sets and evaluate how well the identified equations are able to predict the observed system dynamics. The predicted dynamics will be close to the observed dynamics only if we find a very good approximation of the equations of motion for the systems. For the oscillator and the double oscillator this is relatively easy as the differential equations are linear. For the double pendulum this is very difficult because the equations of motion for the double pendulum are nonlinear and rather complex. Additionally, the double pendulum can be chaotic depending on the starting state, which means that even a very small error in the model can lead to completely different dynamics of simulation and observation after a few seconds.

Table 7.25: General GP parameters used for modelling dynamics of mechanical systems.

Parameter	Value
Population size	1000
Max. generations	1 .. 100
Max. size	$\{5, 10, 15, 20, 25, 30, 35, 40, 45, 50\}$ nodes
Mutation prob.	25%
Crossover prob.	100%
Selection	Tournament (group size 5)
Parameter optimization	10 Levenberg-Marquardt iterations
Fitness	negative MSE
Function set	$F_{\text{Poly}} = \{+, \times\}$
	$F_{\text{ratPoly}} = F_{\text{Poly}} \cup \{\div, \text{AQ}\}$
	$F_{\text{trigPoly}} = F_{\text{Poly}} \cup \{\sin, \cos\}$
	$F_{\text{trigRatPoly}} = F_{\text{ratPoly}} \cup \{\sin, \cos\}$
	$F_{\text{general}} = F_{\text{trigRatPoly}} \cup \{\log, \exp x^2, x^{\frac{1}{2}}, x^{\frac{1}{3}}\}$

7.10.1   Oscillator

The simplest of the four systems is the linear oscillator shown in Figure 7.37. A mass is connected to a spring and slides on a smooth surface. Motion is restricted to one dimension and friction may cause a damped oscillation. Equation (7.27) shows the equation of motion, a linear ODE with three parameters for mass m, spring force coefficient k, and damping coefficient c.

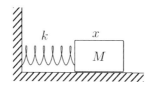

Figure 7.37: The linear damped oscillator.

$$m\ddot{x} + c\dot{x} + kx = 0 \qquad (7.27)$$

This equation models the spring force to be proportional to the elongation of the spring and the frictional force to be proportional to the velocity. Our task is to find an ODE and optimal values for fitting parameters only from observational data. Theoretically, we may find the same equation. Practically, the identified model may be different to better match the observations, because the physical system may have nonlinear effects which are not captured well by the linear model. Figure 7.38 shows the measurements in the dataset `real_linear_a_1.txt` from Schmidt and Lipson (2009) over time.

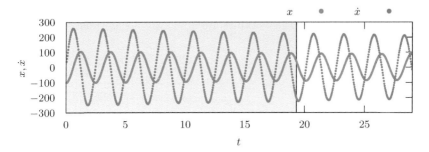

Figure 7.38: Dynamics of state variables in the oscillator dataset. The window with gray background is used as the training set.

Through grid search we found the best results with function set $F_{general}$ and a size limit of 15 nodes with only twenty generations. The runtime for the final GP run with these parameters was just a few seconds and produced the expression tree shown in Figure 7.39. This model (Equation (7.28)) has an NMSE of 3.4×10^{-4} on the training set and 5.5×10^{-4} on the test set.

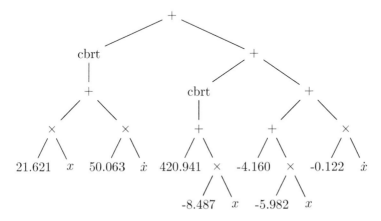

Figure 7.39: The expression tree for the linear oscillator model (Equation (7.28)) identified by GP.

$$M_{\text{SR}} : \ddot{x} = (21.62\,x + 50.06\,\dot{x})^{\frac{1}{3}} - 5.981\,x + (420.9 - 8.487\,x)^{\frac{1}{3}} - 0.1224\,\dot{x} - 4.16 \tag{7.28}$$

Numerically solving the model in Figure 7.39 for the initial values x_0 and $\dot{x}_0$ from the dataset leads to the solution shown in Figure 7.40. The figure clearly shows that the model which was fit to the approximate derivative values leads to a solution which diverges from the measurements in the long term. This divergence can be explained by a small error of the model which adds up over time when simulating the dynamics. The coefficients which fit optimally for the approximate derivative values are obviously not optimal for predicting the long-term dynamics.

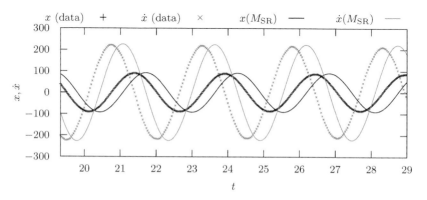

Figure 7.40: The numeric solution of the ODE identified via SR deviates from the measurements in the test set. The initial value for this solution is the first measurement in the training set (not shown).

To improve the model we can simply reoptimize the coefficients to fit the model to the actual dynamics (see Section 6.4). The coefficients identified by SR provide a good starting point for local optimization of coefficients. Equation (7.29) shows the equation after fitting coefficients to the observed x and $\dot{x}$ values.

$$M_{\text{SR opt}} : \ddot{x} = (20.56\,x + 50.41\,\dot{x})^{\frac{1}{3}} - 6.192\,x + (421 - 6.35\,x)^{\frac{1}{3}} - 0.1235\,\dot{x} - 4.337 \tag{7.29}$$

The same procedure can be used to identify the coefficients of the reference model with linear damping:

$$M_{\text{ref.}} : \ddot{x} = -6.23\,x - 0.01616\,\dot{x} \tag{7.30}$$

Figure 7.41 shows the solutions for $M_{\text{SR opt}}$ and $M_{\text{ref.}}$. The coefficients for both models have been fit on the first 300 observations. The solutions of the

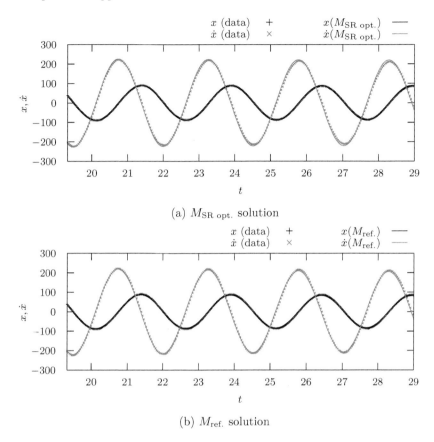

(a) $M_{\text{SR opt.}}$ solution

(b) $M_{\text{ref.}}$ solution

Figure 7.41: The numeric solutions for $M_{\text{SR opt.}}$ and $M_{\text{ref.}}$ match the measurements almost perfectly.

two models are almost indistinguishable and match the measurements almost perfectly.

For this system it is insightful to visualize the two solutions using a vector field plot shown in Figure 7.42 for $M_{\text{SR opt.}}$, and $M_{\text{ref.}}$. In the vector field plots a small difference is visible. While $M_{\text{ref.}}$ has a linear damping behaviour $M_{\text{SR opt.}}$ shows less damping with smaller velocities and smaller displacements.

7.10.2 Pendulum

The next mechanical system we consider is the damped pendulum shown in Figure 7.43. Figure 7.44 shows the measurements in the dataset `real_pend_a_1.txt` from Schmidt and Lipson (2009) over time.

The equation of motion for a damped pendulum with friction losses pro-

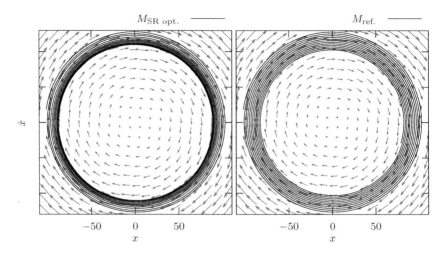

$$M_{\text{SR opt.}} \;\;\text{———}\qquad\qquad M_{\text{ref.}} \;\;\text{———}$$

Figure 7.42: The vector fields and solutions for $M_{\text{SR opt.}}$ (left), and $M_{\text{ref.}}$ (right).

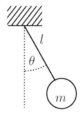

Figure 7.43: Parameters of the pendulum.

portional to the angular velocity can be expressed as shown in Equation (7.31). The parameters for the damping coefficient c, mass m, and length L can be identified by calibrating the model to measurements.

$$m\,\ddot\theta + c\,\dot\theta + \frac{mg}{L}\,\sin\theta = 0 \tag{7.31}$$

As the model requires a nonlinear term it is slightly more difficult for SR, than the linear model for the oscillator. Through grid search we found that function set F_{general} with a size limit of only ten nodes and fourty generations produced good results on average. The model identified with GP in the final run has an NMSE of 5.1×10^{-4} on the training set and 4.4×10^{-4} on the test set. The runtime of the final run with the optimized hyperparameters was only a few seconds. Figure 7.45 shows the expression tree that was found for the angular acceleration $\ddot\theta$ by GP.

$$M_{\text{SR}} : \ddot\theta = -\sin 0.6343\,\dot\theta + \sin\left(0.6293\,\dot\theta - 0.222\,\theta\right) - \sin 1.033\,\theta \tag{7.32}$$

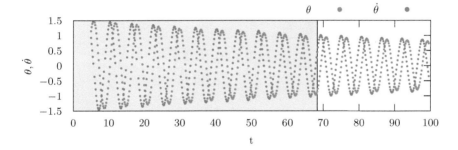

Figure 7.44: Dynamics of state variables in the pendulum dataset. The window with gray background is used as the training set.

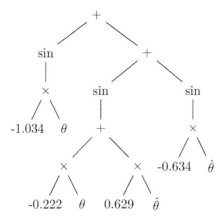

Figure 7.45: The expression tree for $\ddot{\theta}$ of the pendulum identified by GP.

Numerically solving Equation (7.32) for the initial values θ_0 and $\dot{\theta}_0$ in the dataset leads to the solution shown in Figure 7.46. The figure shows that the model which was fit to the approximate derivative values leads to solutions which deviate from the measurements in the test set.

Again, the model can be improved significantly by recalibrating the model coefficients to fit the dynamics on the training set. The SR model with the optimized coefficients $M_{\text{SR opt.}}$ is

$$\ddot{\theta} = -\sin 0.567\,\dot{\theta} + \sin\left(0.561\,\dot{\theta} - 0.281\,\theta\right) - \sin 1.057\,\theta \qquad (7.33)$$

The same procedure can be used to identify the coefficients of the reference model $M_{\text{ref.}}$:

$$\ddot{\theta} = -1.354\sin\theta - 0.0123\,\dot{\theta} \qquad (7.34)$$

Figure 7.47 shows the solutions for $M_{\text{SR opt.}}$ and $M_{\text{ref.}}$. Both models match the measurements almost perfectly even for the forecast in the test set. The solutions for $M_{\text{SR opt.}}$ and $M_{\text{ref.}}$ are almost indistinguishable.

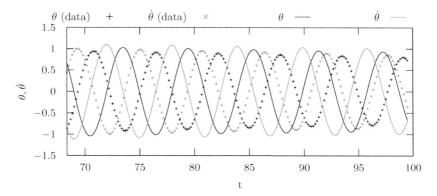

Figure 7.46: The numeric solution for M_{SR} does not fit the data for the test window.

We can see the difference between the two solutions more easily in the vector field plots. Figure 7.48 shows the vector fields for the optimized SR model and the reference model. Again, the SR model predicts slightly less damping for smaller angles than the reference model with linear damping.

7.10.3 Double Oscillator

Finding a model for the double oscillator is slightly more difficult than for the oscillator and the pendulum, because we now have to identify a system of two differential equations to describe the motion of the two masses.

Figure 7.49 shows a sketch of the double oscillator. It has two masses connected to two fixed points with two springs and a third spring connects the two masses to produce a coupled system. A reference model for the equations of motion for the coupled oscillator is shown in Equation (7.36). In this model it is assumed that both masses m are equal, the friction coefficients for both masses are equal, and the three springs are equal. Additionally, we assume that the spring force is linear in the displacement and the friction force is linear in the velocity. Figure 7.50 shows the measurements in the dataset `real_double_linear_a_1.txt` from Schmidt and Lipson (2009).

$$m\,\ddot{x}_1 + c\,\dot{x}_1 + 2\,k\,x_1 - k\,x_2 = 0 \qquad (7.35)$$
$$m\,\ddot{x}_2 + c\,\dot{x}_2 + 2\,k\,x_2 - k\,x_1 = 0 \qquad (7.36)$$

To find the two expressions for the accelerations of the two masses $\ddot{x}_1, \ddot{x}_2$ we can use two independent GP runs using the numerically approximated accelerations as the target variables and the positions and velocities of both masses $x_1, x_2, \dot{x}_1, \dot{x}_2$ as input variables. With grid search we found that function set F_{poly} with a size limit of only ten nodes and only five generations was

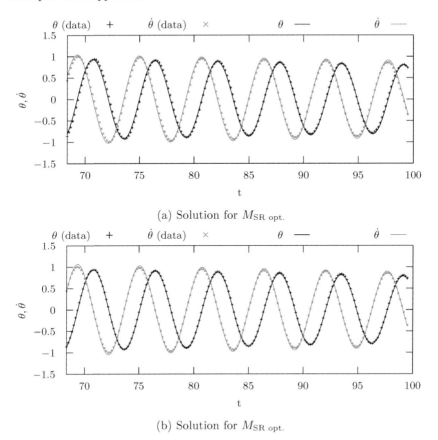

(a) Solution for $M_{\text{SR opt.}}$

(b) Solution for $M_{\text{SR opt.}}$

Figure 7.47: The numeric solutions for $M_{\text{SR opt.}}$ and $M_{\text{ref.}}$ fit the data almost perfectly.

sufficient to produce the best solutions on average. This task is easy for GP because the linear model produces very good solutions and is trivial to find. Both runs for the final model took less than a second of runtime. The model identified for $\ddot{x}_1$ has a NMSE of 4.2×10^{-3} on the training set and 4.4×10^{-3} on the test set. The model for $\ddot{x}_2$ has NMSE of 5.2×10^{-3} on the training set and 7.2×10^{-3} on the test set. The model found by GP is

$$M_{\text{SR}}: \tag{7.37}$$
$$\ddot{x}_1 = 54.92\,x_2 - 61.68\,x_1 - 0.211\,\dot{x}_2 + 0.2155\,\dot{x}_1 + 7295$$
$$\ddot{x}_2 = -25.73\,x_2 + 22.28\,x_1 + 0.07021\,\dot{x}_2 - 0.09034\,\dot{x}_1 - 3000$$

Numerically solving M_{SR} for the initial values $x_1, x_2, \dot{x}_1$, and $\dot{x}_2$ leads to the solution shown in Figure 7.51. As for the oscillator and the pendulum, solving the equations found by GP directly leads to a solution which does not match the measurements in the test set. Here the solution even diverges over

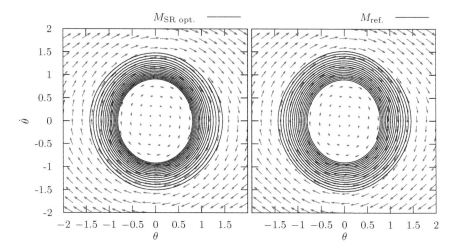

Figure 7.48: The vector fields and solutions for $M_{\text{SR opt.}}$ (left), and M_{ref} (right) show the slightly different damping behaviour captured by the models.

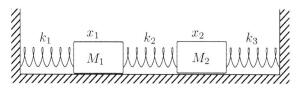

Figure 7.49: Parameters of the double oscillator.

longer time horizons. Therefore, we again have to optimize the coefficients to fit the numerical solution of the model to the observed $x_1, x_2, \dot{x}_1$, and $\dot{x}_2$ values which leads to $M_{\text{SR opt.}}$:

$$\ddot{x}_1 = 77.83\,x_2 - 84.73\,x_1 - 0.2174\,\dot{x}_2 - 0.043\,\dot{x}_1 + 9945$$
$$\ddot{x}_2 = -35.4\,x_2 + 31.96\,x_1 - 0.0171\,\dot{x}_2 + 0.104\,\dot{x}_1 - 4123 \qquad (7.38)$$

The same procedure can be used to identify the coefficients of the reference model $M_{\text{ref.}}$:

$$\ddot{x}_1 = -5.9 \times 10^{-2}\,\dot{x}_1 - 84.82\,x_1 + 78.04\,x_2 + 9950$$
$$\ddot{x}_2 = 3.85 \times 10^{-3}\,\dot{x}_2 + 31.90\,x_1 - 35.39\,x_2 - 4113 \qquad (7.39)$$

We can see that the coefficients for the velocities $\dot{x}_1$ and $\dot{x}_2$ are relatively small, which means that friction only has a minor effect in the observed system. The coefficient identified for $\dot{x}_2$ is positive which is physically not plausible and should be fixed to zero.

Figure 7.52 shows the numerical solutions for $M_{\text{SR opt.}}$ and $M_{\text{ref.}}$. The coefficients for both models have been fit on the first 250 observations. The difference between the two solutions is minimal.

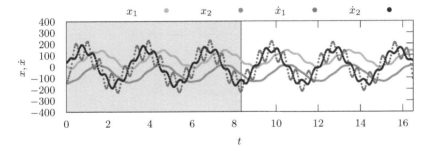

Figure 7.50: Dynamics of state variables for the double oscillator dataset. The window with gray background is used as the training set.

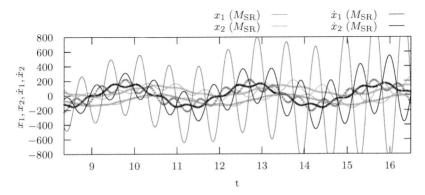

Figure 7.51: The numeric solution of M_{SR} is unstable and diverges in the test partition.

7.10.4 Double Pendulum

The double pendulum is the hardest of the four mechanical systems considered in this section to identify. The reason is that the equations of motion for the double pendulum are nonlinear and relatively complex. Additionally, depending on the initial value the dynamics of the double pendulum can be chaotic. This means that small differences in the starting values of the model can lead to completely different dynamics.

Figure 7.53 shows a sketch of the double pendulum. It has two masses connected through two rods. The state variables are the angles θ_1 and θ_2 and the two angular velocities $\dot{\theta}_1, \dot{\theta}_2$. Our goal is to find the equations of motion to calculate the angular accelerations $\ddot{\theta}_1, \ddot{\theta}_2$ in explicit form.

For small angular displacements θ_1, θ_2 the model M_{approx}:

$$(m_1 + m_2)\, l_1^2\, \ddot{\theta}_1 + m_2\, l_1\, l_2\, \ddot{\theta}_2 + (m_1 + m_2)\, l_1\, g\, \theta_1 = 0 \qquad (7.40)$$
$$m_2\, l_2^2\, \ddot{\theta}_2 + m_2\, l_1\, l_2\, \ddot{\theta}_1 + m_2\, l_2\, g\, \theta_2 = 0$$

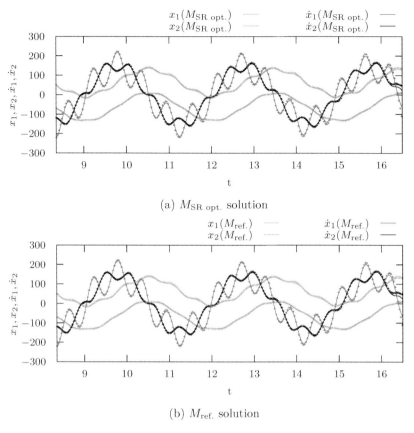

(a) $M_{\text{SR opt.}}$ solution

(b) $M_{\text{ref.}}$ solution

Figure 7.52: The numeric solution for $M_{\text{SR opt.}}$ matches the data almost perfectly, even for the long-term forecast in the test partition. The difference between the optimized SR model and the calibrated reference model $M_{\text{ref.}}$ is minimal.

predicts the dynamics of the double pendulum well (Zill and Cullen, 2005).

In a first step we try to use SR to find a similar model for the small angle case. We use the first part of the dataset `double_pend_h_1.txt` from Schmidt and Lipson (2009) which contains data from a single simulation of the system with constant masses ($m_1 = m_2 = 1$) and lengths ($l_1 = l_2 = 1$) shown in Figure 7.54. This dataset is split into a training set (two-thirds) and a test set (last third).

We use the hyperparameters from Table 7.25 and again use a grid search to select the function set, size limit, and number of generations. The results of the grid search show that function set F_{general} with a size limit of just 15 nodes and only thirty generations is sufficient to produce good results on average. The runtime of the final run with the best hyperparameters for the two equations

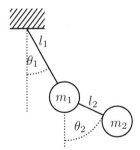

Figure 7.53: Parameters of the double pendulum.

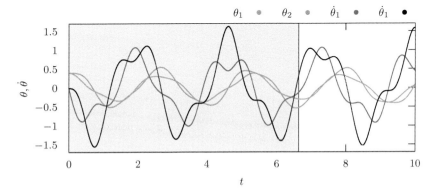

Figure 7.54: Dynamics of state variables for the double pendulum with small angles and velocities. The gray rectangle marks the training window.

for $\ddot{\theta}_1$ and $\ddot{\theta}_2$ is just a few seconds. The fitting parameters capture the combined effects of masses and lengths but we cannot identify these physical parameters from the coefficients. The SR model M_{SR}:

$$\ddot{\theta}_1 = 0.005127\, \dot{\theta}_1 + \sin\left(5.18\, \theta_2 - 3.391\, \theta_1\right)$$
$$+ \sin\left(3.993\, \theta_2 - 6.496\, \theta_1\right) + 2.902\, \theta_2 - 11.97\, \theta_1$$

$$\ddot{\theta}_2 = -7.876\, \sin\left(\sin\left(2.593\, \theta_2 - 1.623\, \theta_1\right) - 2.22\, \theta_1\right) - \sin 2.11\, \theta_2 - 7.748\, \theta_1$$

has a NMSE of 1.5×10^{-4} for $\ddot{\theta}_1$ on the training set and 1.4×10^{-4} on the test set. For $\ddot{\theta}_2$ the prediction is even better with NMSE of 2.31×10^{-6} on the training set and 2.58×10^{-6} on the test set.

Trying to solve M_{SR} for the initial values $\theta_1(0), \theta_2(0), \dot{\theta}_1(0)$, and $\dot{\theta}_2(0)$ from the training set fails because its solution quickly diverges. After optimizing the coefficients to minimize the SSE for the observed angles θ_1, θ_2 the model

$M_{\text{SR opt.}}$ is

$$\ddot{\theta}_1 = 4.502 \times 10^{-3}\,\dot{\theta}_1 + \sin(-8.74\,\theta_2 + 13.64\,\theta_1)$$
$$+ \sin(5.937\,\theta_2 - 11\,\theta_1) + 12.33\,\theta_2 - 21.97\,\theta_1$$
$$\ddot{\theta}_2 = -13.29\sin(\sin(1.543\,\theta_2 - 6.362 \times 10^{-3}\,\theta_1) - 1.552\,\theta_1)$$
$$- \sin 1.392\,\theta_2 + 0.1\,\theta_1 \tag{7.41}$$

The full reference model used to generate the data for `double_pend_h_1` by Schmidt and Lipson (2009) is shown in Equation (7.42). The model assumes $m_1 = m_2 = 1$, and $l_1 = l_2 = 1$, that the rods connecting the masses are rigid and have no mass and there are no friction losses leading to damping.

$$M_{\text{ref.}}: \tag{7.42}$$
$$g = 9.81$$
$$c = \frac{1}{2g}$$
$$\ddot{\theta}_1 = \frac{1}{c + c\sin(\theta_1 - \theta_2)^2}\left(\frac{1}{2}\sin(\theta_2)\cos(\theta_1 - \theta_2)\right.$$
$$\left. - \sin(\theta_1 - \theta_2)(c\,\dot{\theta}_1^2\cos(\theta_1 - \theta_2) + c\,\dot{\theta}_2^2) - \sin(\theta_1)\right)$$
$$\ddot{\theta}_2 = \frac{1}{c + c\sin(\theta_1 - \theta_2)^2}\left(2\,c\,\dot{\theta}_1^2\sin(\theta_1 - \theta_2) + \sin(\theta_1)\cos(\theta_1 - \theta_2)\right.$$
$$\left. + c\,\dot{\theta}_2^2\sin(\theta_1 - \theta_2)\cos(\theta_1 - \theta_2) - \sin(\theta_2)\right)$$

Figure 7.55 shows the numerical solutions for $M_{\text{SR opt.}}$ where the model coefficients have been reoptimized on the training window which contains the first 866 observations of the simulated dynamics. The plot shows only the test window. The optimized SR model identified by GP accurately captures the dynamics for the small-angle case which is demonstrated by the accuracy of the predictions to the target points.

Next, we can try to find a better model which is also able to predict the dynamics for large angles and angular velocities. Because of the potential difficulties of identifying the equations of motion for a real double pendulum we will use only the second part from the synthetic dataset `double_pend_h_1.txt` from Schmidt and Lipson (2009). The dataset is generated using $M_{\text{ref.}}$ (Equation (7.42)). We use this dataset completely as a test set and generate nine additional datasets using the same parameters as well as the initial values shown in Table 7.26. Each dataset spans a simulation of ten seconds. These additional simulated datasets are helpful for GP to find a model that works for different initial conditions. We use the first six datasets for training and three datasets together with the second part of `double_pend_h_1.txt` for testing.

Finding the equations of motion for the double pendulum is much harder than for the other three mechanical systems. Therefore, we increase the population size to 10,000 individuals and adjust the tournament group size to seven

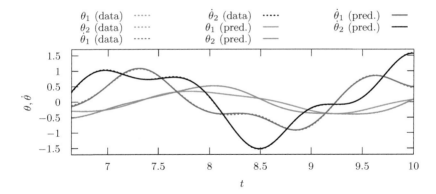

Figure 7.55: The numeric solution of $M_{\text{SR opt.}}$ for this initial state of the double pendulum with small angles produces an almost perfect prediction.

Table 7.26: Initial values for the datasets used for training and test for the double pendulum model.

Dataset	Initial values $(\theta_1, \theta_2, \dot{\theta}_1, \dot{\theta}_2)$
Train-1	(1.0, 0.0, 0.0, 0.0)
Train-2	(2.0, 0.0, 0.0, 0.0)
Train-3	(3.0, 0.0, 0.0, 0.0)
Train-4	(1.0, 1.0, 0.0, 0.0)
Train-5	(1.0, 2.0, 0.0, 0.0)
Train-6	(1.0, 3.0, 0.0, 0.0)
Test-1	(1.0, −1.0, 0.0, 0.0)
Test-2	(1.0, −2.0, 0.0, 0.0)
Test-3	(1.0, −3.0, 0.0, 0.0)
double_pend_h_1.txt	$(\frac{3}{4}\pi,\ \pi,\ -0.01,\ -0.01)$

to reach a similar selection pressure. For the other hyperparameters we use the values shown in Table 7.25.

We use the function set $\{+, -, \times, \div, \sin, \cos\}$ and the terminal set $\{\theta_1, \theta_2, \dot{\theta}_1, \dot{\theta}_2\} \cup \{c_i\}_{i=1..100}$, whereby we initalize the coefficients c_i uniformly at random in the interval $[-20, 20]$. The coefficients are optimized in each fitness evaluation. The maximum size limit is set to fifty nodes and we use 100 generations. We run thirty repetitions and select the best models on the training set. The two expressions found by GP have a NMSE $< 10^{-6}$ for both $\ddot{\theta}_1$ and $\ddot{\theta}_2$ on the training set. The runtime for these two runs was five hours and 13 hours (HeuristicLab, single-threaded).

The SR model found by GP with coefficients optimized to minimize SSE to θ_1 and θ_2 is $M_{\text{SR opt.}}$:

$$\ddot{\theta}_1 = -\sin\left(0.675 \sin\left(1.848 \sin\sin\left(\theta_2 - \theta_1\right)\right)\right)$$

$$\left(0.01 - 0.801\,\dot{\theta}_2^2 + \cos\left(\theta_2 - \theta_1\right)\left(-0.8\,\dot{\theta}_1^2 - 7.89 \sin\left(\theta_1 + 14.13\right)\right)\right)$$

$$+ 9.811 \sin\left(\theta_1 + 15.7\right) + 0.002824$$

$$\ddot{\theta}_2 = \sin\left(\sin\sin\sin\sin\left(\theta_2 - \theta_1\right) + \sin\sin\left(\theta_2 - \theta_1\right)\right)$$

$$\left(-0.5015 \cos\left(\theta_2 - \theta_1\right)\dot{\theta}_2^2 - 1.003\,\dot{\theta}_1^2 - 9.829 \cos\theta_1\right) + 0.001031$$

Figure 7.56 shows the solutions for the initial values for the six training sets and Figure 7.57 shows the solutions for the test sets. The different solution result from different initial values but the same model with the same coefficients was used for all solutions. The plots show that the predictions are accurate for the datasets with periodic dynamics even for initial values used for the test sets. However, for the datasets with chaotic dynamics, the prediction of the model deviates from the simulated data after a few seconds. For some initial values the dynamics are chaotic, and even small inaccuracies in the model would lead to large prediction errors in the long term. Considering this difficulty, it is striking that GP was able to find a model which at least predicts the non-chaotic dynamics almost perfectly, and is able to predict a few seconds of the chaotic dynamics.

In summary, with GP-based SR we were able to identify systems of differential equations for simple mechanical systems purely from data. For the simple systems, we found models that predicted the dynamics equally well as the reference models. For the double pendulum, we found an approximation that predicts small-angle dynamics well but fails to predict chaotic dynamics with larger velocities.

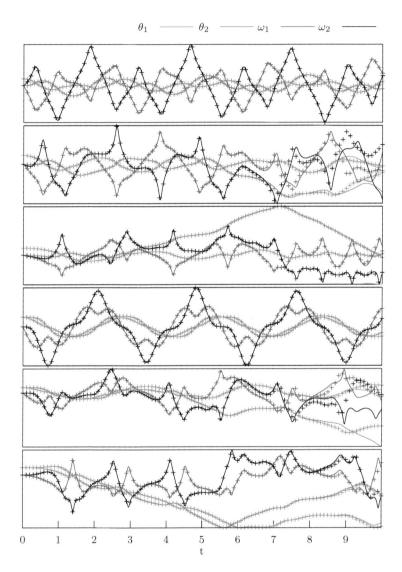

Figure 7.56: Solutions for $M_{\text{SR opt.}}$ for the six training sets.

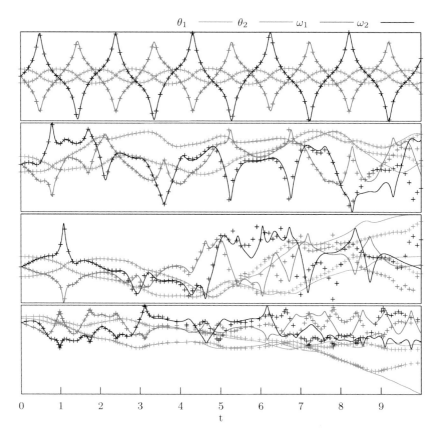

Figure 7.57: Solutions for $M_{\text{SR opt.}}$ for the four test sets.

7.11 Conclusions

The examples given in this chapter demonstrate the versatility of SR with GP. For each of the modelling tasks we used a slightly different solution approach and demonstrated several of the customizations or advanced techniques that are discussed in Chapter 6. In these examples we used dimensionless parameters, factor variables, multi-objective optimization, gradient-based local optimization of coefficients, differential equations, autoregressive models, classification, cross-validation, model-selection, model simplification, prediction intervals, and several other techniques.

We have carefully selected a set of diverse examples from multiple domains of science and engineering and hope that our step-by-step descriptions are a helpful guide for experimenting with SR yourself. A good way to familiarize yourself with SR is to try to reproduce the results for the same datasets which are available online, or alternatively you can follow the same workflow to solve similar tasks using your own datasets.

8

Conclusion

Symbolic regression (SR) by genetic programming (GP) is without any doubt one of the most powerful methods for interpretable machine learning (ML). It is highly flexible with respect to basic functions it uses as well as the complexity of the models it produces.

With this book, our aim was to summarize the basic concepts as well as advanced concepts of SR and GP as well as exemplary applications not only for computer scientists and data scientists, but also for domain experts from various fields. We explained basic concepts of ML, summarized SR basics as well as evolutionary computation and GP. Then we took a deeper dive into validation, inspection, and postprocessing of models, summarized advanced concepts such as coefficients optimization, multi-objective optimization, and the identification of differential equations. Finally, we described a variation of applications in which we have successfully used explainable ML by GP for solving regression, classification, and time series tasks. In the appendix, we give an overview of existing software resources and benchmarks that are relevant in the context of SR.

8.1 Unique Selling Points of Symbolic Regression

We are convinced that SR can make significant contributions to artificial intelligence and ML, both in theoretical aspects as well as in practical applications.

There is an increasing interest in explainability in AI and ML. Especially when the goal is to find out not only what, but even more why something happens, i.e., when we discuss not only correlations, but rather causalities, then the explainability and interpretability of models are essential. An example for this is root cause analysis in the area of predictive maintenance, where failure causes have to be explained and interpreted, and the origins have to be identified. Explainability and interpretability are also key elements for the establishment of human-centered AI, as it is much easier, more intuitive and comprehensible for humans to interact with AI systems if they understand the internal processes and can interpret the resulting models. More than other currently known ML methods, SR allows identifying complex nonlinear relationships in data in a form that can still be explained and interpreted.

SR enables us to integrate and hybridize data- and knowledge-based approaches rather than viewing them as mutually exclusive. When the mechanisms of a system are partially known, then this can be used e.g., via shape constraints or predefined partial model structure to improve models. In this way, models can be identified that represent observational data well and at the same time are consistent with process knowledge. Thus, spurious correlations can be avoided, as they can occur easily with sparse datasets, and models with better interpolation and extrapolation capabilities can be found.

Using SR we are able to specifically tailor the properties and the power of the hypothesis space to the requirements of the task at hand. By determining which mathematical operators appear to be useful for describing interrelationships in a particular domain, certain mathematical operators may or may not be allowed as nonterminal symbols. This enables SR to provide sufficiently complex hypothesis spaces and thus to avoid underfitting. Equally important is the avoidance of too powerful hypothesis spaces, which in turn may lead to overfitting.

Another advantage of SR is the genericity of its representation. This allows a comparatively simple transformation of models into all common programming languages. One may even use SR models in spreadsheet software. The models can be used easily within many commercial simulation software tools used in engineering.

Finally, using SR we can deploy models also on comparatively limited hardware. This is a critical advantage for example in the context of industrial control systems, which initiates possibilities in the field of edge computing. SR models can be evaluated close to real time and can for example be used as virual sensors for model predictive control.

8.2   Limitations and Caveats

In this book we gave many examples where SR is well suited and we mentioned multiple advantages of SR. However, there are also cases where it is appropriate to be careful when using SR. This book would not be complete if we would not give some hints when you should refrain from using SR.

SR is not suited well for analyzing unstructured data. This is especially the case when it comes to text, images, videos, and audio data. Other methods, especially artificial neural networks, work with huge amounts of input signals (such as pixels), compress and transform these data, and eventually, for instance, generate text, classify images, or recognize objects in images. Of course, in theory SR could also be used to build extremely huge model structures and achieve the same calculations, but this would be misguided as SR was developed particularly for the purpose of generating shorter interpretable models.

The results of machine learning should not be interpreted to assume cau-

sation where there is only correlation. Correlation or rather relationships are a good indicator, but not sufficient to establish causal effects. A nice short formula produced by SR that establishes a nonlinear relationship between a "dependent" variable and multiple "independent" variables might be convincing to suspect a causal relationship. However, as with all statistical modelling, SR can only establish correlations. Nevertheless, due to the inherent interpretability of the models learned in this way, SR is more likely to reveal pseudo-correlations than, for example, methods that do not allow for interpretation. However, as we have shown above SR is a great tool for causal modelling when it is used together with a well-designed experiment to gather measurements that can be used for causal modelling and with prior knowledge (e.g., physics-based constraints). Especially, time series models including lags, or in the form of differential equations, where we explicitly distinguish between controlled variables and state variables, are inherently causal models.

SR models should be used conscientiously and reflectively when facing "unknown" or surprising correlations. Again, this is a general issue for all predictive models methods. It is tempting to use ML methods including SR for rather unguided data mining to find unsuspected correlations. This seems to be especially common for predicting financial time series or in biology in the form of genomic wide association studies. The danger of using SR for this task is that SR evaluates many different correlations including even interactions between variables and nonlinear transformations of variables. As a result there is a high probability to find spurious correlations as the number of observations is usually by far not enough relative to the number of potential relationships that are considered by the algorithm.

It is not easy to use SR when the model has to include many input variables. SR implicitly requires a selection of the most relevant variables from a larger set of potentially relevant variables. An SR model may only reference a few variables, assuming that we are not using aggregation functions that map a vector of input variables to a scalar value. Therefore, for modelling tasks where a large number of features all affect the dependent variable equally, SR is not an ideal choice. For example for tasks where categorical input variables with many possible values have to be one-hot-encoded.

In conclusion, we are convinced that SR will play an even more important role in the future as it is a flexible and generic technique that implements explainable AI. As artificial intelligence is becoming more and more important for so many applications, it is becoming even more important for us to understand why models react in the way they react – and also why they are identified in the way they are identified.

9

Appendix

SR is rapidly becoming more popular especially in the natural sciences. As a consequence, many SR implementations are developed with different features and aims. Most of the implementations are maintained by researchers in academia and available open source. Additionally, a few commercial systems are also available today. As a consequence of the rapid development of the field, any list of software implementations is outdated as soon as it is published. New projects are started frequently and at the same time many projects are discontinued. In the following list we have included a few larger software systems and frameworks that have been around for a while and seem to be well-supported. All of the listed systems use different forms of GP for SR. Additionally to the mentioned software systems several good software modules exist that implement non-evolutionary algorithms for SR which we have mentioned in Section 6.6.

9.1 Benchmarks

The symbolic regression benchmark project (SRBench) (Orzechowski et al., 2018; La Cava et al., 2021) is an ongoing effort to improve comparability and reproducibility among SR methods, by introducing an open-source, curated collection of benchmark problems accessible via a Scikit-learn compatible API.

The datasets included in the benchmark suite have been gathered and indexed by the Computational Genetics Lab at the University of Pennsylvania (Olson et al., 2017). The initiative helps ML researchers assess the performance of their methods in the areas of regression and classification by providing a solid baseline to benchmark against and a standardized way to measure performance. At the same time, it helps practitioners choose the method that best suits their needs by providing a continuously updated, live ranking of methods.

In recent years, the Python language and its scientific ecosystem have become the de-facto standard in data science. The integration of new methods with SRBench can be achieved with a minimal amount of wrapper code necessary to provide a Python interface for interoperability with the benchmarking framework. Most other programming languages provide bindings for Python language making this task easily manageable.

A large number of software implementations of algorithms for SR have been included into SRBench and can be easily installed, tested, and compared via a common Python interface. For many algorithms the curated fork integrated in SRBench is the easiest way to run and test them. As of the end of 2023 the following algorithms are available:[1]

- Age-Fitness Pareto Optimization (Schmidt and Lipson, 2010)

- Age-Fitness Pareto Optimization with Co-evolved Fitness Predictors (Schmidt and Lipson, 2010)

- AI-Feynman 2.0 (Udrescu et al., 2020)

- Bayesian Symbolic Regression (Jin et al., 2020)

- Deep Symbolic Regression (Petersen et al., 2021)

- Fast Function Extraction (FFX) (McConaghy, 2011)

- Feature Engineering Automation Tool (FEAT) (La Cava et al., 2019)

- Epsilon-Lexicase Selection (La Cava et al., 2016)

- GP-based Gene-pool Optimal Mixing Evolutionary Algorithm (GP-GOMEA) (Virgolin et al., 2017)

- gplearn `https://gplearn.readthedocs.io`

- Interaction-Transformation Evolutionary Algorithm (de Franca and Aldeia, 2021)

- Multiple Regression GP (Arnaldo et al., 2014)

- Operon (Burlacu et al., 2020)

- Semantic Backpropagation GP (Virgolin et al., 2019)

Homepage: `https://cavalab.org/srbench/`

9.2   Open-source Software for Genetic Programming

9.2.1   HeuristicLab

HeuristicLab (Wagner et al., 2014) is one of the most general and mature frameworks for metaheuristic optimization. It features a flexible abstraction

---

[1]Source: `https://github.com/cavalab/srbench`

layer between problem domains (e.g., combinatorial optimization, regression, classification) and solving methods (e.g., GP, evolution strategies, tabu search, etc.) which enables a high degree of modularity and code reuse. Optimization algorithms can be applied to different kinds of problem encodings, for instance binary strings, real vectors, syntax trees, each having their own set of encoding-specific operators defined (e.g., interpreters, evaluators, genetic operators, neighborhood-based operators, and so on). The software is under active development using the Microsoft .NET framework and can be used as a library (Linux and Windows compatible) or as a full standalone desktop application (Windows only) with extended analysis and visualization capabilities.
Homepage: `https://github.com/heal-research/HeuristicLab`

9.2.2 Operon

Operon (Burlacu et al., 2020) is a C++ implementation of GP-based SR with a focus on memory efficiency and parallel execution. The C++ library is complemented by a Python module called PyOperon which exposes its functionality to Python and additionally provides a *scikit-learn* estimator interface. Many algorithmic ideas and genetic operators in Operon are inspired by Heuristic-Lab. The Operon library uses a linear tree encoding with postfix indexing and evolves models using a fine-grained concurrency scheme where every recombination event takes place in its own logical thread. It supports single- and multi-objective optimization, automatic differentiation of expressions and nonlinear least-squares optimization of model parameters. Operon is part of SRBench and achieves good results on many datasets (La Cava et al., 2021).
Homepage: `https://github.com/heal-research/operon`

9.2.3 PySR

The development of PySR (Cranmer, 2023) was motivated by the need to discover interpretable equations in science, which must provide insight into the nature of the problem. PySR implements multi-population GP where each population evolves asynchronously, following the classic structure of selection, mutation, and recombination. Mutation is further enhanced with a temperature parameter similar to simulated annealing, such that mutations are accepted or rejected with some probability based on the change in fitness after mutation. After these steps, simplification and parameter optimization may also be applied.

In order to promote simplicity, the algorithm applies parsimony pressure by adding a term penalising the expression complexity to the loss function. Expressions are considered in the context of the entire population and their complexity score is adapted based on their frequency over a moving time window. To facilitate working with noisy data, PySR optionally performs a denoising step and allows individual weights for each data point. Other important features include custom operators, custom loss functions, expression constraints, and automatic feature selection.

The PySR framework is implemented in Julia under the library name SYMBOLICREGRESSION.JL. Since Julia is a just-in-time compiled (JIT) language, it can achieve high execution speed comparable to languages such as C or C++, while also allowing user-defined operators and loss functions to be JIT compiled or even fused into efficient SIMD-enabled kernels. PySR exposes a Python frontend to the pure Julia library backend and follows the API specified by the *scikit-learn* machine learning library (Pedregosa et al., 2011).
Homepage: `https://astroautomata.com/PySR`

9.2.4 DEAP

DEAP (Distributed Evolutionary Algorithms in Python) (Fortin et al., 2012) is a Python framework supporting a large number of single- and multi-objective evolutionary algorithms. It aims to provide transparent mechanisms for the user to define their own types and operators for rapid prototyping and testing of ideas. A number of common optimization algorithms such as standard GP, NSGA-II, or SPEA are provided out of the box. It's main strength comes from being supported by a large number of Python scientific computing libraries (NumPy, SciPy, Pandas, Matplotlib, etc.) as well as parallelization frameworks such as Dask or Scoop. It's limitations are inherent to the Python programming language which being an interpreted language will not offer the same level of performance as other frameworks.
Homepage: `https://deap.readthedocs.io`

9.2.5 FEAT

FEAT (Feature Engineering Automation Tool) (La Cava et al., 2019) is a multi-objective evolutionary approach where each model is a linear combination of a set of evolved features. FEAT uses a tree-like model representation where the nodes are various Boolean and continuous-valued functions, including neural activation functions such as tanh, sigmoid, logit, or relu nodes. Model parameters are encoded into the edges between nodes and are optimized using stochastic gradient descent. The models are evolved using usual genetic operators (e.g., crossover, mutation) and a model complexity measure is used as a secondary objective to guide the search. FEAT is geared primarily towards solving SR problems, however symbolic classification is possible by passing the input of the model through a logistic function.
Homepage: `https://cavalab.org/feat`

9.2.6 ECJ

ECJ (Evolutionary Computation in Java) (Scott and Luke, 2019) is a library for evolutionary computation that provides a large amount of algorithms such as genetic algorithms, GP, evolution strategies, differential evolution, or particle swarm optimization. Geared towards experts and research scientists, ECJ is one of the most popular and most used research libraries in the field of GP and has been under active development for the last twenty years.

Among all the supported methods, GP for SR is one of the framework's strongest suits. As opposed to other frameworks which require an amount of programming to set up, ECJ experiments can be entirely configured using parameter files formatted in a simple text format. The configuration system follows a folder hierarchy such that sophisticated configurations can be achieved by the hierarchical nesting of parameter files. Although this increases the steepness of the learning curve for using the framework, it offers very extensive options for preparing and running experiments.

The framework can be additionally extended by Java programmers by making use of its object-oriented design and available abstractions in order to prototype and implement new methods and algorithms.

Homepage: `https://cs.gmu.edu/~eclab/projects/ecj`

9.2.7 GPTIPS

GPTIPS (Searson, 2015) is an open-source MATLAB-based framework for SR implementing a GP variant called multigene GP (MGGP), where each model is a weighted linear combination of GP trees – called "genes" in the context of MGGP. The main assumption behind MGGP is that a linear combination of scaled tree outputs can capture nonlinear behavor more effectively than a single scaled tree output and lends itself well to automatic post-run model simplification.

GPTIPS further constrains the MGGP models to certain depth and complexity limits and uses Pareto tournament selection to ensure the evolution of compact models. Each model can contain a random number of genes between one and a user-defined maximum. However, a large number of genes risks overfitting the data by adding terms that contribute very little to the model output. The crossover operator is adapted to work with multigene models and can move entire genes between models or swap individual subtrees within the genes.

Due to the increased complexity of the representation, GPTIPS has a higher computational cost than standard GP. Mainly, the calculation of gene weighting coefficients using least-squares fitting carries a heavy computational burden and takes up a large proportion of the algorithm's runtime.

Homepage: `https://www.gptips.org`

9.3 Commercial Software for Genetic Programming

9.3.1 Evolved Analytics DataModeler

DataModeler (Kotanchek et al., 2013) is a proprietary extension for the Mathematica^{TM} software that provides SR capabilities. Internally, DataModeler implements a multi-objective GP algorithm optimizing for accuracy and parsimony at the same time using a Pareto dominance approach. The algorithm considers the knee of the Pareto front as the region containing models that have the best tradeoff between accuracy and complexity and seeks to create an ensemble of models.

The software includes advanced analysis and postprocessing capabilities for identifying the most relevant variables in the data and for producing "metavariables" (defined as transforms of the original variables) which can provide additional insight into the modelling problem.

Homepage: `https://evolved-analytics.com`

9.3.2 Eureqa Formulize

Eureqa (Schmidt and Lipson, 2009) is a closed-source software suite for the "reverse engineering of dynamical systems", that can learn symbolic equations describing nonlinear dynamical systems from time series data. An educated guess about its inner workings can be made from previous research articles and the following algorithmic aspects can be identified:[2]

- A directed acyclic graph encoding is used instead of the traditional syntax tree, in order to achieve a better compromise between regression performance, space, and computational complexity (Schmidt and Lipson, 2007).

- Co-evolution of fitness predictors (Schmidt and Lipson, 2008a) is used to speed up fitness assignment, by optimizing a second population of training sample indices to maximize the ability to separate good equations from bad ones.

- Age-fitness Pareto optimization (Schmidt and Lipson, 2010) is used to provide an influx of new genetic material in each generation in the form of new offspring which are protected from competing against older, more fit individuals until they reach a certain level of maturity.

---

[2]Due to the closed-source nature of the software, we cannot guarantee the accuracy of this description.

Bibliography

M. Affenzeller, S. Winkler, S. Wagner, and A. Beham. *Genetic Algorithms and Genetic Programming – Modern Concepts and Practical Applications*, volume 6 of *Numerical Insights*. CRC Press, Chapman & Hall, 2009. ISBN 978-1-58488-629-7.

M. Affenzeller, S. M. Winkler, B. Burlacu, G. Kronberger, M. Kommenda, and S. Wagner. Dynamic observation of genotypic and phenotypic diversity for different symbolic regression GP variants. In *Proceedings of the Genetic and Evolutionary Computation Conference Companion*, GECCO '17, pages 1553–1558, Berlin, Germany, July 2017. ACM. doi: 10.1145/3067695.3082530.

K. Ahnert and M. Abel. Numerical differentiation of experimental data: local versus global methods. *Computer Physics Communications*, 177(10):764–774, Nov. 2007. doi: 10.1016/j.cpc.2007.03.009.

H. Akaike. A new look at the statistical model identification. *IEEE Transactions on Automatic Control*, 19(6):716–723, 1974. doi: 10.1109/TAC.1974.1100705.

G. Alefeld and G. Mayer. Interval analysis: theory and applications. *Journal of Computational and Applied Mathematics*, 121(1):421–464, 2000. ISSN 0377-0427. doi: 10.1016/S0377-0427(00)00342-3.

D. M. Allen. The relationship between variable selection and data augmentation and a method for prediction. *Technometrics*, 16(1):125–127, 1974. doi: 10.1080/00401706.1974.10489157.

L. Alonso and R. Schott. *Random Generation of Trees: Random Generators in Computer Science*. Kluwer Academic Publishers, 1995. ISBN 079239528X.

L. Altenberg. The evolution of evolvability in genetic programming. In K. E. Kinnear, Jr., editor, *Advances in Genetic Programming*, chapter 3, pages 47–74. MIT Press, 1994a.

L. Altenberg. Emergent Phenomena in Genetic Programming. In *Evolutionary Programming Proceedings of the Third Annual Conference*, pages 233–241. World Scientific Publishing, 1994b.

G. L. Andriole, E. D. Crawford, R. L. Grubb III, S. S. Buys, D. Chia, T. R. Church, M. N. Fouad, E. P. Gelmann, P. A. Kvale, D. J. Reding, et al.

Bibliography

Mortality results from a randomized prostate-cancer screening trial. *New England Journal of Medicine*, 360(13):1310–1319, 2009.

I. Arnaldo, K. Krawiec, and U.-M. O'Reilly. Multiple regression genetic programming. In *Proceedings of the 2014 Annual Conference on Genetic and Evolutionary Computation*, GECCO '14, pages 879–886. ACM, 2014. ISBN 9781450326629. doi: 10.1145/2576768.2598291.

M. Z. Asadzadeh, H.-P. Gänser, and M. Mücke. Symbolic regression based hybrid semiparametric modelling of processes: an example case of a bending process. *Applications in Engineering Science*, 6:100049, June 2021. doi: 10.1016/j.apples.2021.100049.

T. Back. Selective pressure in evolutionary algorithms: a characterization of selection mechanisms. In *Proceedings of the First IEEE Conference on Evolutionary Computation. IEEE World Congress on Computational Intelligence*, volume 1, pages 57–62, 1994. doi: 10.1109/ICEC.1994.350042.

J. Bader and E. Zitzler. HypE: an algorithm for fast hypervolume-based many-objective optimization. *Evolutionary Computation*, 19(1):45–76, 2011. doi: 10.1162/EVCO_a_00009.

F. Baeta, J. Correia, T. Martins, and P. Machado. TensorGP – genetic programming engine in TensorFlow. In P. Castillo and J. Jimenez-Laredo, editors, *24th International Conference, EvoApplications 2021*, volume 12694 of *Lecture Notes in Computer Sciences*, pages 763–778. Springer Verlag, Apr. 2021. doi: 10.1007/978-3-030-72699-7_48.

W. Böhm and A. Geyer-Schulz. Exact uniform initialization for genetic programming. In R. K. Belew and M. Vose, editors, *Foundations of Genetic Algorithms IV*, pages 379–407. Morgan Kaufmann, Aug. 1996. ISBN 1-55860-460-X.

W. Banzhaf, E. Goodman, L. Sheneman, L. Trujillo, and B. Worzel, editors. *Genetic Programming Theory and Practice XVII*. Springer International Publishing, 2020. doi: 10.1007/978-3-030-39958-0.

W. Banzhaf, L. Trujillo, S. Winkler, and B. Worzel, editors. *Genetic Programming Theory and Practice XVIII*. Springer Nature Singapore, 2022. doi: 10.1007/978-981-16-8113-4.

D. Barber. *Bayesian Reasoning and Machine Learning*. Cambridge University Press, 2012. doi: 10.1017/CBO9780511804779.

D. Bartlett, H. Desmond, and P. Ferreira. Priors for symbolic regression. In *Proceedings of the Companion Conference on Genetic and Evolutionary Computation*. ACM, July 2023a. doi: 10.1145/3583133.3596327.

D. J. Bartlett, H. Desmond, and P. G. Ferreira. Exhaustive symbolic regression. *IEEE Transactions on Evolutionary Computation*, pages 1–1, 2023b. doi: 10.1109/tevc.2023.3280250.

D. M. Bates and D. G. Watts. *Nonlinear Regression Analysis and Its Applications*. John Wiley & Sons, Inc., 1988. doi: 10.1002/9780470316757.

L. Beadle and C. Johnson. Semantically driven crossover in genetic programming. In J. Wang, editor, *Proceedings of the IEEE World Congress on Computational Intelligence*, pages 111–116, Hong Kong, June 2008. IEEE Computational Intelligence Society, IEEE Press. doi: 10.1109/CEC.2008.4630784.

L. Beadle and C. G. Johnson. Semantic analysis of program initialisation in genetic programming. *Genetic Programming and Evolvable Machines*, 10(3): 307–337, Sep. 2009. ISSN 1389-2576. doi: 10.1007/s10710-009-9082-5.

Y. Bengio and Y. LeCun. Scaling learning algorithms towards AI. In L. Bottou, O. Chapelle, D. DeCoste, and J. Weston, editors, *Large Scale Kernel Machines*. MIT Press, 2007.

M. Berz and G. Hoffstätter. Computation and application of Taylor polynomials with interval remainder bounds. *Reliable Computing*, 4(1):83–97, Feb. 1998. ISSN 1573-1340. doi: 10.1023/A:1009958918582.

M. Berz and K. Makino. Verified integration of ODEs and flows using differential algebraic methods on high-order Taylor models. *Reliable Computing*, 4 (4):361–369, Nov. 1998. ISSN 1573-1340. doi: 10.1023/A:1024467732637.

M. Berz, K. Makino, and J. Hoefkens. Verified integration of dynamics in the solar system. *Nonlinear Analysis: Theory, Methods & Applications*, 47 (1):179–190, 2001. ISSN 0362-546X. doi: 10.1016/S0362-546X(01)00167-5. Proceedings of the Third World Congress of Nonlinear Analysts.

C. M. Bishop. *Neural Networks for Pattern Recognition*. Oxford University Press, Nov. 1995. ISBN 9780198538493. doi: 10.1093/oso/9780198538493. 001.0001.

C. M. Bishop. *Pattern Recognition and Machine Learning*. Springer New York, NY, 1st edition, 2006. ISBN 978-0-387-31073-2. doi: 10.1007/ 978-0-387-45528-0.

R. Bivand. Revisiting the Boston data set – Changing the units of observation affects estimated willingness to pay for clean air. *REGION*, 4(1):109–127, May 2017. doi: 10.18335/region.v4i1.107.

T. Blickle and L. Thiele. A comparison of selection schemes used in evolutionary algorithms. *Evolutionary Computation*, 4(4):361–394, Winter 1996. ISSN 1063-6560. doi: 10.1162/evco.1996.4.4.361.

G. Bomarito, T. Townsend, K. Stewart, K. Esham, J. Emery, and J. Hochhalter. Development of interpretable, data-driven plasticity models with symbolic regression. *Computers and Structures*, 252:106557, Aug. 2021. doi: 10.1016/j.compstruc.2021.106557.

J. Brandstetter, D. Worrall, and M. Welling. Message passing neural PDE solvers. In *ICLR 2022*, Apr. 2022. doi: 10.48550/arXiv.2202.03376.

L. Breiman. Random forests. *Machine Learning*, 45(1):5–32, 2001.

L. Breiman, J. Friedman, C. J. Stone, and R. A. Ohlshen. *Classification and Regression Trees*. Chapman & Hall, 1984.

F. V. V. Breugel, J. N. Kutz, and B. W. Brunton. Numerical differentiation of noisy data: a unifying multi-objective optimization framework. *IEEE Access*, 8:196865–196877, 2020. doi: 10.1109/access.2020.3034077.

S. L. Brunton and J. N. Kutz. *Data-Driven Science and Engineering: Machine Learning, Dynamical Systems, and Control*. Cambridge University Press, 2nd edition, 2022. doi: 10.1017/9781009089517.

S. Bubeck. Convex optimization: algorithms and complexity. *Foundations and Trends in Machine Learning*, 8:231–357, 2015.

E. K. Burke, S. Gustafson, and G. Kendall. Diversity in genetic programming: an analysis of measures and correlation with fitness. *IEEE Transactions on Evolutionary Computation*, 8(1):47–62, Feb. 2004. ISSN 1089-778X. doi: 10.1109/TEVC.2003.819263.

A. R. Burks and W. F. Punch. *Genetic Programming Theory and Practice XIV*, chapter An Investigation of Hybrid Structural and Behavioral Diversity Methods in Genetic Programming, pages 19–34. Springer International Publishing, 2018. ISBN 978-3-319-97088-2. doi: 10.1007/978-3-319-97088-2_2.

B. Burlacu, M. Affenzeller, G. Kronberger, and M. Kommenda. Online diversity control in symbolic regression via a fast hash-based tree similarity measure. In *2019 IEEE Congress on Evolutionary Computation (CEC)*, pages 2175–2182, 2019. doi: 10.1109/CEC.2019.8790162.

B. Burlacu, G. Kronberger, and M. Kommenda. Operon C++: an efficient genetic programming framework for symbolic regression. In *Proceedings of the 2020 Genetic and Evolutionary Computation Conference Companion*, GECCO '20, pages 1562–1570. ACM, July 2020. doi: 10.1145/3377929.3398099.

K. Burnham and D. Anderson. *Model Selection and Multimodel Inference: A Practical Information-Theoretic Approach*. Springer New York, 2003. ISBN 9780387953649.

Y. Cao, S. Li, L. Petzold, and R. Serban. Adjoint sensitivity analysis for differential-algebraic equations: the adjoint DAE system and its numerical solution. *SIAM Journal on Scientific Computing*, 24(3):1076–1089, Jan. 2003. doi: 10.1137/s1064827501380630.

M. Caracotsios and W. E. Stewart. Sensitivity analysis of initial value problems with mixed ODEs and algebraic equations. *Computers and Chemical Engineering*, 9(4):359–365, Jan. 1985. doi: 10.1016/0098-1354(85)85014-6.

M. F. Causley. The gamma function via interpolation. *Numerical Algorithms*, 90(2):687–707, 2022. ISSN 1572-9265. doi: 10.1007/s11075-021-01204-8.

R. Chartrand. Numerical differentiation of noisy, nonsmooth data. *ISRN Applied Mathematics*, 2011:1–11, May 2011. doi: 10.5402/2011/164564.

R. T. Q. Chen, Y. Rubanova, J. Bettencourt, and D. K. Duvenaud. Neural ordinary differential equations. In S. Bengio, H. Wallach, H. Larochelle, K. Grauman, N. Cesa-Bianchi, and R. Garnett, editors, *Advances in Neural Information Processing Systems*, volume 31. Curran Associates, Inc., 2018.

T. Chen and C. Guestrin. XGBoost: a scalable tree boosting system. In *Proceedings of the 22nd ACM SIGKDD International Conference on Knowledge Discovery and Data Mining*, KDD '16, pages 785–794. ACM, 2016. ISBN 9781450342322. doi: 10.1145/2939672.2939785.

W. S. Cleveland. Robust locally weighted regression and smoothing scatterplots. *Journal of the American Statistical Association*, 74(368):829–836, 1979.

C. A. Coello Coello. Theoretical and numerical constraint-handling techniques used with evolutionary algorithms: a survey of the state of the art. *Computer Methods in Applied Mechanics and Engineering*, 191(11):1245–1287, 2002. ISSN 0045-7825. doi: 10.1016/S0045-7825(01)00323-1.

T. M. Cover and J. A. Thomas. *Elements of Information Theory*. John Wiley & Sons, Inc., 2005. doi: 10.1002/047174882x.

N. L. Cramer. A representation for the adaptive generation of simple sequential programs. In *Proceedings of the 1st International Conference on Genetic Algorithms*, pages 183–187, USA, 1985. L. Erlbaum Associates Inc. ISBN 0805804269.

M. Cranmer. Interpretable machine learning for science with PySR and SymbolicRegression.jl, 2023. arxiv preprint 2305.01582.

M. Črepinšek, S.-H. Liu, and M. Mernik. Exploration and exploitation in evolutionary algorithms: a survey. *ACM Computing Surveys*, 45(3):35:1–35:3, July 2013. ISSN 0360-0300. doi: 10.1145/2480741.2480752.

M. Curmei and G. Hall. Shape-constrained regression using sum of squares polynomials. *Operations Research*, Sep. 2023. doi: 10.1287/opre.2021.0383.

L. H. de Figueiredo and J. Stolfi. Affine arithmetic: concepts and applications. *Numerical Algorithms*, 37(1):147–158, Dec. 2004. ISSN 1572-9265. doi: 10.1023/B:NUMA.0000049462.70970.b6.

F. O. de Franca and G. S. I. Aldeia. Interaction–transformation evolutionary algorithm for symbolic regression. *Evolutionary Computation*, 29(3):367–390, Sep. 2021. ISSN 1063-6560. doi: 10.1162/evco_a_00285.

F. O. de Franca and G. Kronberger. Reducing overparameterization of symbolic regression models with equality saturation. In *Proceedings of the Genetic and Evolutionary Computation Conference*, GECCO '23, pages 1064–1072. ACM, 2023. ISBN 9798400701191. doi: 10.1145/3583131.3590346.

B. M. de Silva, D. M. Higdon, S. L. Brunton, and J. N. Kutz. Discovery of physics from data: universal laws and discrepancies. *Frontiers in Artificial Intelligence*, 3:25, 2020. ISSN 2624-8212. doi: 10.3389/frai.2020.00025.

K. Deb and H. Jain. An evolutionary many-objective optimization algorithm using reference-point-based nondominated sorting approach, part I: solving problems with box constraints. *IEEE Transactions on Evolutionary Computation*, 18:577–601, 2014.

K. Deb, A. Pratap, S. Agarwal, and T. Meyarivan. A fast and elitist multiobjective genetic algorithm: NSGA-II. *IEEE Transactions on Evolutionary Computation*, 6(2):182–197, Apr. 2002. ISSN 1941-0026. doi: 10.1109/4235.996017.

O. Dekel, S. Shalev-Shwartz, and Y. Singer. Smooth epsilon-insensitive regression by loss symmetrization. *Journal of Machine Learning Research*, 6: 711–741, Jan. 2005.

D. Dickmanns, J. Schmidhuber, and A. Winklhofer. Der genetische Algorithmus: Eine Implementierung in Prolog. Fortgeschrittenenpraktikum, Institut für Informatik, Technische Universität München, 1987.

S. Dignum and R. Poli. Generalisation of the limiting distribution of program sizes in tree-based genetic programming and analysis of its effects on bloat. In *GECCO '07: Proceedings of the 9th Annual Conference on Genetic and Evolutionary Computation*, volume 2, pages 1588–1595, London, July 2007. ACM Press. doi: 10.1145/1276958.1277277.

N. R. Draper and H. Smith. *Applied Regression Analysis*. Wiley, 1998.

R. O. Duda, P. E. Hart, and D. G. Stork. *Pattern Classification*. Wiley Interscience, 2nd edition, 2000. ISBN 0-471-05669-3.

M. Duffy, P. McGowan, and W. Gallagher. Cancer invasion and metastasis: changing views. *The Journal of Pathology*, 214(3):283–293, 2008. doi: 10.1002/path.2282.

L. J. Eshelman and J. D. Schaffer. Preventing premature convergence in genetic algorithms by preventing incest. In *Proceedings of the 4th International Conference on Genetic Algorithms (ICGA '91)*, pages 115–122. Morgan Kaufmann, 1991.

T. Fawcett. An introduction to ROC analysis. *Pattern Recognition Letters*, 27 (8):861–874, 2006. ISSN 0167-8655. doi: 10.1016/j.patrec.2005.10.010.

C. Ferreira. Gene expression programming in problem solving. In R. Roy, M. Köppen, S. Ovaska, T. Furuhashi, and F. Hoffmann, editors, *Soft Computing and Industry: Recent Applications*, pages 635–653. Springer London, 2002. ISBN 978-1-4471-0123-9. doi: 10.1007/978-1-4471-0123-9_54.

A. Fisher, C. Rudin, and F. Dominici. All models are wrong, but many are useful: learning a variable's importance by studying an entire class of prediction models simultaneously. *Journal of Machine Learning Research*, 20(177):1–81, 2019.

D. B. Fogel and L. J. Fogel. An introduction to evolutionary programming. In J.-M. Alliot, E. Lutton, E. Ronald, M. Schoenauer, and D. Snyers, editors, *Artificial Evolution*, volume 1063 of *Lecture Notes in Computer Science*, pages 21–33. Springer Berlin Heidelberg, 1996. ISBN 978-3-540-61108-0. doi: 10.1007/3-540-61108-8_28.

F.-A. Fortin, F.-M. De Rainville, M.-A. Gardner, M. Parizeau, and C. Gagné. DEAP: evolutionary algorithms made easy. *Journal of Machine Learning Research*, 13:2171–2175, July 2012.

J. H. Friedman. Multivariate adaptive regression splines. *The Annals of Statistics*, 19(1):1–67, Mar. 1991.

J. H. Friedman. Greedy function approximation: a gradient boosting machine. *The Annals of Statistics*, 29(5):1189–1232, Oct. 2001. doi: 10.1214/aos/1013203451.

J. H. Friedman, E. Grosse, and W. Stuetzle. Multidimensional additive spline approximation. *SIAM Journal on Scientific and Statistical Computing*, 4(2): 291–301, 1983. doi: 10.1137/0904023.

E. Galvan-Lopez, J. McDermott, M. O'Neill, and A. Brabazon. Defining locality as a problem difficulty measure in genetic programming. *Genetic Programming and Evolvable Machines*, 12(4):365–401, Dec. 2012. ISSN 1389-2576. doi: 10.1007/s10710-011-9136-3.

E. Gardeñes, M. Á. Sainz, L. Jorba, R. Calm, R. Estela, H. Mielgo, and A. Trepat. Model intervals. *Reliable Computing*, 7(2):77–111, Apr. 2001. doi: 10.1023/A:1011465930178.

S. Gaucel, M. Keijzer, E. Lutton, and A. Tonda. Learning dynamical systems using standard symbolic regression. In *Lecture Notes in Computer Science*, pages 25–36. Springer Berlin Heidelberg, 2014. doi: 10.1007/978-3-662-44303-3_3.

A. Gelman, J. B. Carlin, H. S. Stern, D. Dunson, A. Vehtari, and D. Rubin. *Bayesian Data Analysis*. Chapman and Hall/CRC, 3rd edition, 2013.

J. Gerritsma, R. Onnink, and A. Versluis. Geometry, resistance and stability of the delft systematic yacht hull series. *International Shipbuilding Progress*, 28:276–297, 1981.

A. Geyer–Schulz. *Fuzzy Rule-Based Expert Systems and Genetic Machine Learning*, volume 3 of *Studies in Fuzziness and Soft Computing*. Physica-Verlag, 2nd revised edition, 1996.

O. W. Gilley and R. Pace. On the Harrison and Rubinfeld data. *Journal of Environmental Economics and Management*, 31(3):403–405, 1996. ISSN 0095-0696. doi: 10.1006/jeem.1996.0052.

K. Goebel, B. Saha, A. Saxena, J. R. Celaya, and J. P. Christophersen. Prognostics in battery health management. *IEEE Instrumentation & Measurement Magazine*, 11(4):33–40, 2008. doi: 10.1109/MIM.2008.4579269.

P. Gold and S. O. Freedman. Demonstration of tumor-specific antigens in human colonic carcinomata by immunological tolerance and absorption techniques. *Journal of Experimental Medicine*, 121(3):439–462, Mar. 1965. ISSN 0022-1007. doi: 10.1084/jem.121.3.439.

D. E. Goldberg. *Genetic Algorithms in Search, Optimization and Machine Learning*. Addison-Wesley Longman Publishing Co., Inc., 1st edition, 1989. ISBN 0201157675.

I. Goodfellow, Y. Bengio, and A. Courville. *Deep Learning*. Adaptive Computation and Machine Learning. MIT Press, 2016. ISBN 9780262035613.

P. D. Grünwald. *The Minimum Description Length Principle*. The MIT Press, 2007. doi: 10.7551/mitpress/4643.001.0001.

A. Griewank and A. Walther. *Evaluating Derivatives: Principles and Techniques of Algorithmic Differentiation*. Society for Industrial and Applied Mathematics, 2nd edition, Jan. 2008. doi: 10.1137/1.9780898717761.

G. Grimmett and D. Stirzaker. *Probability and Random Processes*. Oxford University Press, 2020.

J. Han, M. Kamber, and J. Pei. *Data Mining: Concepts and Techniques*. Morgan Kaufmann, 2011.

W. Handley and P. Lemos. Quantifying dimensionality: Bayesian cosmological model complexities. *Physical Review D*, 100:023512, July 2019. doi: 10.1103/PhysRevD.100.023512.

E. R. Hansen. Sharpness in interval computations. *Reliable Computing*, 3(1): 17–29, Feb. 1997. ISSN 1573-1340. doi: 10.1023/A:1009917818868.

M. H. Hansen and B. Yu. Model selection and the principle of minimum description length. *Journal of the American Statistical Association*, 96(454): 746–774, 2001. doi: 10.1198/016214501753168398.

N. Hansen, S. D. Müller, and P. Koumoutsakos. Reducing the time complexity of the derandomized evolution strategy with covariance matrix adaptation (CMA-ES). *Evolutionary Computation*, 11(1):1–18, 2003. doi: 10.1162/106365603321828970.

D. Harrison and D. Rubinfeld. Hedonic housing prices and demand for clean air. *Journal of Environmental Economics and Management*, 5(1):81–102, 1978. doi: 10.1016/0095-0696(78)90006-2.

T. Hastie, R. Tibshirani, and J. Friedman. *The Elements of Statistical Learning*. Springer Series in Statistics. Springer New York Inc., 2001.

N. Heckert, J. Filliben, C. Croarkin, B. Hembree, W. Guthrie, P. Tobias, and J. Prinz. Handbook 151: NIST/SEMATECH e-handbook of statistical methods, Nov. 2002.

T. Helmuth and L. Spector. Evolving a digital multiplier with the PushGP genetic programming system. In *GECCO '13 Companion: Proceeding of the 15th Annual Conference on Genetic and Evolutionary Computation Conference Companion*, pages 1627–1634, Amsterdam, The Netherlands, July 2013. ACM. doi: 10.1145/2464576.2466814.

A. Hernandez, A. Balasubramanian, F. Yuan, S. A. M. Mason, and T. Mueller. Fast, accurate, and transferable many-body interatomic potentials by symbolic regression. *npj Computational Materials*, 5(1):112, Nov. 2019. ISSN 2057-3960. doi: 10.1038/s41524-019-0249-1.

K. Hingee and M. Hutter. Equivalence of probabilistic tournament and polynomial ranking selection. In *2008 IEEE Congress on Evolutionary Computation (IEEE World Congress on Computational Intelligence)*, pages 564–571, 2008.

N. X. Hoai, R. I. McKay, and H. A. Abbass. Tree adjoining grammars, language bias, and genetic programming. In C. Ryan, T. Soule, M. Keijzer, E. Tsang, R. Poli, and E. Costa, editors, *Genetic Programming, Proceedings of EuroGP'2003*, volume 2610 of *Lecture Notes in Computer Sciences*, pages 335–344, Essex, Apr. 2003. Springer-Verlag. ISBN 3-540-00971-X. doi: 10.1007/3-540-36599-0_31.

A. E. Hoerl and R. W. Kennard. Ridge regression: biased estimation for nonorthogonal problems. *Technometrics*, 12(1):55–67, Feb. 1970.

M. D. Hoffman and A. Gelman. The No-U-Turn sampler: adaptively setting path lengths in Hamiltonian Monte Carlo. *Journal of Machine Learning Research*, 15(1):1593–1623, 2014.

J. H. Holland. *Adaptation in Natural and Artificial Systems*. The University of Michigan Press, 1st edition, 1975.

P. Holmes and P. J. Barclay. Functional languages on linear chromosomes. In *Proceedings of the 1st Annual Conference on Genetic Programming*, pages 427–427, Cambridge, MA, USA, 1996. MIT Press. ISBN 0-262-61127-9.

G. S. Hornby. ALPS: the age-layered population structure for reducing the problem of premature convergence. In *Proceedings of the 8th Annual Conference on Genetic and Evolutionary Computation*, pages 815–822. ACM, 2006.

D. N. T. How, M. A. Hannan, M. S. Hossain Lipu, and P. J. Ker. State of charge estimation for lithium-ion batteries using model-based and data-driven methods: a review. *IEEE Access*, 7:136116–136136, 2019. doi: 10.1109/ACCESS.2019.2942213.

T. Hu and W. Banzhaf. Quantitative analysis of evolvability using vertex centralities in phenotype network. In *Proceedings of the Genetic and Evolutionary Computation Conference 2016*, GECCO '16, pages 733–740. ACM, 2016a. ISBN 9781450342063. doi: 10.1145/2908812.2908940.

T. Hu and W. Banzhaf. Neutrality, robustness and evolvability in genetic programming. In R. Riolo, W. P. Worzel, and M. Kotanchek, editors, *Genetic Programming Theory and Practice XIV*, Genetic and Evolutionary Computation. Springer, 2016b.

T. Hu, M. Tomassini, and W. Banzhaf. A network perspective on genotype-phenotype mapping in genetic programming. *Genetic Programming and Evolvable Machines*, 21(3):375–397, Sep. 2020. ISSN 1389-2576. doi: 10.1007/s10710-020-09379-0. Special Issue: Highlights of Genetic Programming 2019 Events.

P. J. Huber. Robust estimation of a location parameter. *The Annals of Mathematical Statistics*, 35(1):73 – 101, 1964. doi: 10.1214/aoms/1177703732.

H. Iba. Inference of differential equation models by genetic programming. *Information Sciences*, 178(23):4453–4468, Dec. 2008. doi: 10.1016/j.ins.2008.07.029.

H. Jain and K. Deb. An evolutionary many-objective optimization algorithm using reference-point based nondominated sorting approach, part II: handling constraints and extending to an adaptive approach. *IEEE Transactions on Evolutionary Computation*, 18:602–622, 2014.

Y. Jin, W. Fu, J. Kang, J. Guo, and J. Guo. Bayesian symbolic regression, 2020. arxiv preprint 1910.08892.

E. Kabliman, A. H. Kolody, J. Kronsteiner, M. Kommenda, and G. Kronberger. Application of symbolic regression for constitutive modeling of plastic deformation. *Applications in Engineering Science*, 6:100052, June 2021. doi: 10.1016/j.apples.2021.100052.

L. Kammerer, G. Kronberger, B. Burlacu, S. M. Winkler, M. Kommenda, and M. Affenzeller. Symbolic regression by exhaustive search: reducing the search space using syntactical constraints and efficient semantic structure deduplication. In *Genetic Programming Theory and Practice XVII*, pages 79–99. Springer International Publishing, 2020. doi: 10.1007/978-3-030-39958-0_5.

M. Keijzer. Improving symbolic regression with interval arithmetic and linear scaling. In C. Ryan, T. Soule, M. Keijzer, E. Tsang, R. Poli, and E. Costa, editors, *Genetic Programming, Proceedings of EuroGP'2003*, volume 2610 of *Lecture Notes in Computer Sciences*, pages 70–82, Essex, Apr. 2003. Springer-Verlag. ISBN 3-540-00971-X. doi: 10.1007/3-540-36599-0_7.

M. Keijzer. Scaled symbolic regression. *Genetic Programming and Evolvable Machines*, 5(3):259–269, 2004. ISSN 1389-2576.

M. Keijzer and J. Foster. Crossover bias in genetic programming. In *Genetic Programming*, pages 33–44. Springer, 2007.

A. Keshaviah, S. Dellapasqua, N. Rotmensz, J. Lindtner, D. Crivellari, J. Collins, M. Colleoni, B. Thürlimann, C. Mendiola, S. Aebi, et al. Ca15-3 and alkaline phosphatase as predictors for breast cancer recurrence: a combined analysis of seven international breast cancer study group trials. *Annals of Oncology*, 18(4):701–708, 2007.

D. Kinzett, M. Johnston, and M. Zhang. How online simplification affects building blocks in genetic programming. In *GECCO '09: Proceedings of the 11th Annual Conference on Genetic and Evolutionary Computation*, pages 979–986, Montreal, July 2009. ACM. doi: 10.1145/1569901.1570035.

S. Kirkpatrick, C. D. Gelatt, and M. P. Vecchi. Optimization by simulated annealing. *Science*, 220(4598):671–680, May 1983. doi: 10.1126/science.220.4598.671.

M. Kommenda, G. Kronberger, M. Affenzeller, S. Winkler, and B. Burlacu. Evolving simple symbolic regression models by multi-objective genetic programming. In R. Riolo, W. P. Worzel, M. Kotanchek, and A. Kordon, editors, *Genetic Programming Theory and Practice XIII*, Genetic and Evolutionary Computation, pages 1–19, Ann Arbor, USA, May 2015. Springer. doi: 10.1007/978-3-319-34223-8_1.

M. Kommenda, B. Burlacu, G. Kronberger, and M. Affenzeller. Parameter identification for symbolic regression using nonlinear least squares. *Genetic Programming and Evolvable Machines*, 21(3):471–501, Sep. 2020. ISSN 1573-7632. doi: 10.1007/s10710-019-09371-3.

A. Kordon. Evolutionary computation in the chemical industry. In T. Yu, L. Davis, C. Baydar, and R. Roy, editors, *Evolutionary Computation in Practice*, pages 245–262. Springer Berlin Heidelberg, 2008. ISBN 978-3-540-75771-9. doi: 10.1007/978-3-540-75771-9_11.

A. Kordon, F. Castillo, G. Smits, and M. Kotanchek. Application issues of genetic programming in industry. In *Genetic Programming Theory and Practice III*, pages 241–258. Springer US, 2006. ISBN 978-0-387-28111-7. doi: 10.1007/0-387-28111-8_16.

M. E. Kotanchek, E. Vladislavleva, and G. Smits. Symbolic regression is not enough: it takes a village to raise a model. In *Genetic Programming Theory and Practice X*, pages 187–203. Springer New York, 2013. ISBN 978-1-4614-6846-2. doi: 10.1007/978-1-4614-6846-2_13.

J. R. Koza. *Genetic Programming: On the Programming of Computers by Means of Natural Selection*. MIT Press, 1992. ISBN 0-262-11170-5.

J. R. Koza. *Genetic Programming II: Automatic Discovery of Reusable Programs*. MIT Press, May 1994. ISBN 0-262-11189-6.

J. R. Koza, D. Andre, F. H. Bennett III, and M. Keane. *Genetic Programming III: Darwinian Invention and Problem Solving*. Morgan Kaufmann, Apr. 1999. ISBN 1-55860-543-6.

J. R. Koza, M. A. Keane, M. J. Streeter, W. Mydlowec, J. Yu, and G. Lanza. *Genetic Programming IV: Routine Human-Competitive Machine Intelligence*. Kluwer Academic Publishers, 2003. ISBN 1-4020-7446-8.

K. Krawiec. *Behavioral Program Synthesis with Genetic Programming*, volume 618 of *Studies in Computational Intelligence*. Springer International Publishing, 2016. ISBN 978-3-319-27563-5. doi: 10.1007/978-3-319-27565-9.

K. Krawiec and T. Pawlak. Approximating geometric crossover by semantic backpropagation. In *Proceedings of the 15th Annual Conference on Genetic and Evolutionary Computation*, GECCO '13, pages 941–948. ACM, 2013a. ISBN 9781450319638. doi: 10.1145/2463372.2463483.

K. Krawiec and T. Pawlak. Locally geometric semantic crossover: a study on the roles of semantics and homology in recombination operators. *Genetic Programming and Evolvable Machines*, 14(1):31–63, Mar. 2013b. ISSN 1573-7632. doi: 10.1007/s10710-012-9172-7.

G. Kronberger, M. Kommenda, A. Promberger, and F. Nickel. Predicting friction system performance with symbolic regression and genetic programming with factor variables. In *Proceedings of the Genetic and Evolutionary Computation Conference.* ACM, July 2018. doi: 10.1145/3205455.3205522.

G. Kronberger, L. Kammerer, and M. Kommenda. Identification of dynamical systems using symbolic regression. In *Computer Aided Systems Theory – EUROCAST 2019*, pages 370–377. Springer International Publishing, 2020. doi: 10.1007/978-3-030-45093-9_45.

G. Kronberger, F. O. de Franca, B. Burlacu, C. Haider, and M. Kommenda. Shape-constrained symbolic regression – improving extrapolation with prior knowledge. *Evolutionary Computation*, 30(1):75–98, Mar. 2022. ISSN 1063-6560. doi: 10.1162/evco_a_00294.

W. H. Kruskal and W. A. Wallis. Use of ranks in one-criterion variance analysis. *Journal of the American Statistical Association*, 47(260):583–621, 1952.

W. La Cava, L. Spector, and K. Danai. Epsilon-lexicase selection for regression. In T. Friedrich, editor, *GECCO '16: Proceedings of the 2016 Annual Conference on Genetic and Evolutionary Computation*, pages 741–748, Denver, USA, July 2016. ACM. doi: 10.1145/2908812.2908898.

W. La Cava, T. R. Singh, J. Taggart, S. Suri, and J. H. Moore. Learning concise representations for regression by evolving networks of trees. In *International Conference on Learning Representations*, ICLR, 2019.

W. La Cava, P. Orzechowski, B. Burlacu, F. de Franca, M. Virgolin, Y. Jin, M. Kommenda, and J. Moore. Contemporary symbolic regression methods and their relative performance. In J. Vanschoren and S. Yeung, editors, *Proceedings of the Neural Information Processing Systems Track on Datasets and Benchmarks*, volume 1, 2021.

R.-S. Lai, C.-C. Chen, P.-C. Lee, and J.-Y. Lu. Evaluation of Cytokeratin 19 fragment (CYFRA 21-1) as a tumor marker in malignant pleural effusion. *Japanese Journal of Clinical Oncology*, 29(9):421–424, Sep. 1999. ISSN 0368-2811. doi: 10.1093/jjco/29.9.421.

W. B. Langdon. Size fair and homologous tree genetic programming crossovers for tree genetic programming. *Genetic Programming and Evolvable Machines*, 1:95–119, 2000. doi: 10.1023/A:1010024515191.

W. B. Langdon. Incremental evaluation in genetic programming. In *Genetic Programming: 24th European Conference, EuroGP 2021, Held as Part of EvoStar 2021*, pages 229–246. Springer-Verlag, 2021. ISBN 978-3-030-72811-3. doi: 10.1007/978-3-030-72812-0_15.

W. B. Langdon, T. Soule, R. Poli, and J. A. Foster. The evolution of size and shape. In L. Spector, W. B. Langdon, U.-M. O'Reilly, and P. J. Angeline, editors, *Advances in Genetic Programming 3*, chapter 8, pages 163–190. MIT Press, June 1999. ISBN 0-262-19423-6.

S. Lipovetsky and M. Conklin. Analysis of regression in game theory approach. *Applied Stochastic Models in Business and Industry*, 17(4):319–330, 2001. doi: 10.1002/asmb.446.

D. C. Liu and J. Nocedal. On the limited memory BFGS method for large scale optimization. *Mathematical Programming*, 45(1-3):503–528, Aug. 1989. doi: 10.1007/bf01589116.

L. Ljung. *System Identification: Theory for the User.* Pearson Education, 2nd edition, Dec. 1998. ISBN 9780132441933.

S. Luke. Two fast tree-creation algorithms for genetic programming. *IEEE Transactions on Evolutionary Computation*, 4(3):274–283, 2000.

S. Luke and L. Panait. A survey and comparison of tree generation algorithms. In *GECCO'01: Proceedings of the 3rd Annual Conference on Genetic and Evolutionary Computation*, pages 81–88, July 2001.

S. M. Lundberg and S.-I. Lee. A unified approach to interpreting model predictions. In I. Guyon, U. V. Luxburg, S. Bengio, H. Wallach, R. Fergus, S. Vishwanathan, and R. Garnett, editors, *Advances in Neural Information Processing Systems*, volume 30. Curran Associates, Inc., 2017.

S. M. Lundberg, G. Erion, H. Chen, A. DeGrave, J. M. Prutkin, B. Nair, R. Katz, J. Himmelfarb, N. Bansal, and S.-I. Lee. From local explanations to global understanding with explainable AI for trees. *Nature Machine Intelligence*, 2(1):2522–5839, 2020.

D. J. C. MacKay. *Information Theory, Inference and Learning Algorithms.* Cambridge University Press, 2003. ISBN 9780521642989.

H. Mühlenbein and D. Schlierkamp-Voosen. Predictive models for the breeder genetic algorithm – I. continuous parameter optimization. *Evolutionary Computation*, 1(1):25–49, Mar. 1993. doi: 10.1162/evco.1993.1.1.25.

H. Majeed and C. Ryan. Using context-aware crossover to improve the performance of GP. In M. Keijzer, M. Cattolico, D. Arnold, V. Babovic, C. Blum, P. Bosman, M. V. Butz, C. Coello Coello, D. Dasgupta, S. G. Ficici, J. Foster, A. Hernandez-Aguirre, G. Hornby, H. Lipson, P. McMinn, J. Moore, G. Raidl, F. Rothlauf, C. Ryan, and D. Thierens, editors, *GECCO 2006: Proceedings of the 8th Annual Conference on Genetic and Evolutionary Computation*, volume 1, pages 847–854, Seattle, Washington, USA, July 2006. ACM Press. ISBN 1-59593-186-4. doi: 10.1145/1143997.1144146.

G. Martius and C. H. Lampert. Extrapolation and learning equations. In *5th International Conference on Learning Representations, ICLR 2018-Workshop Track Proceedings*, 2017. doi: 10.48550/arXiv.1610.02995.

T. McConaghy. FFX fast, scalable, deterministic symbolic regression technology. In *Genetic Programming Theory and Practice IX*, pages 235–260. Springer, 2011.

P. McCullagh and J. A. Nelder. *Generalized Linear Models*. CRC Press, 2nd edition, 1989. doi: 10.1201/9780203753736.

N. F. McPhee and N. J. Hopper. Analysis of genetic diversity through population history. In *Proceedings of the Genetic and Evolutionary Computation Conference*, volume 2, pages 1112–1120. Morgan Kaufmann, 1999. ISBN 1-55860-611-4.

N. F. McPhee and J. D. Miller. Accurate replication in genetic programming. In L. J. Eshelman, editor, *Genetic Algorithms: Proceedings of the Sixth International Conference (ICGA95)*, pages 303–309, Pittsburgh, PA, USA, July 1995. Morgan Kaufmann. ISBN 1-55860-370-0.

A. Meurer, C. P. Smith, M. Paprocki, O. Čertík, S. B. Kirpichev, M. Rocklin, A. Kumar, S. Ivanov, J. K. Moore, S. Singh, T. Rathnayake, S. Vig, B. E. Granger, R. P. Muller, F. Bonazzi, H. Gupta, S. Vats, F. Johansson, F. Pedregosa, M. J. Curry, A. R. Terrel, Š. Roučka, A. Saboo, I. Fernando, S. Kulal, R. Cimrman, and A. Scopatz. SymPy: symbolic computing in python. *PeerJ Computer Science*, 3:e103, Jan. 2017. ISSN 2376-5992. doi: 10.7717/peerj-cs.103.

A. Miller. *Subset Selection in Regression*. Chapman and Hall/CRC, Apr. 2002. doi: 10.1201/9781420035933.

B. L. Miller and D. E. Goldberg. Genetic algorithms, tournament selection, and the effects of noise. *Complex Systems*, 9(3):193–212, 1995.

J. F. Miller. An empirical study of the efficiency of learning Boolean functions using a Cartesian Genetic Programming approach. In W. Banzhaf, J. Daida, A. E. Eiben, M. H. Garzon, V. Honavar, M. Jakiela, and R. E. Smith, editors, *Proceedings of the Genetic and Evolutionary Computation Conference*, volume 2, pages 1135–1142, Orlando, Florida, USA, July 1999. Morgan Kaufmann. ISBN 1-55860-611-4.

G. J. Mizejewski. Alpha-fetoprotein structure and function: relevance to isoforms, epitopes, and conformational variants. *Experimental Biology and Medicine*, 226:377–408, 2001.

C. Molnar. *Interpretable Machine Learning*. Lulu.com, 2nd edition, 2020.

D. J. Montana. Strongly typed genetic programming. BBN Technical Report #7866, Bolt Beranek and Newman, Inc., 10 Moulton Street, Cambridge, MA 02138, USA, May 1993.

D. C. Montgomery, C. L. Jennings, and M. Kulahci. *Introduction to Time Series Analysis and Forecasting*. John Wiley & Sons, Inc., 2007.

J. J. Moré, B. S. Garbow, and K. E. Hillstrom. User guide for MINPACK-1. Technical report, CM-P00068642, 1980.

A. Moraglio, K. Krawiec, and C. G. Johnson. Geometric semantic genetic programming. In C. A. C. Coello, V. Cutello, K. Deb, S. Forrest, G. Nicosia, and M. Pavone, editors, *Parallel Problem Solving from Nature - PPSN XII*, volume 7491 of *Lecture Notes in Computer Science*, pages 21–31. Springer, Berlin, Heidelberg, 2012. ISBN 978-3-642-32937-1. doi: 10.1007/978-3-642-32937-1_3.

P. Moscato. On evolution, search, optimization, genetic algorithms and martial arts: towards memetic algorithms. Technical Report Caltech Concurrent Computation Program, 158-79, California Institute of Technology, 1989.

P. Moscato, H. Sun, and M. N. Haque. Analytic continued fractions for regression: a memetic algorithm approach. *Expert Systems with Applications*, 179:115018, 2021. ISSN 0957-4174. doi: 10.1016/j.eswa.2021.115018.

K. G. Murty and S. N. Kabadi. Some NP-complete problems in quadratic and nonlinear programming. *Mathematical Programming*, 39(2):117–129, June 1987. doi: 10.1007/bf02592948.

S. N. and K. Deb. Multiobjective optimization using nondominated sorting in genetic algorithms. *Evolutionary Computation*, 2(3):221–248, Sep. 1994. ISSN 1063-6560. doi: 10.1162/evco.1994.2.3.221.

F. Neri, C. Cotta, and P. Moscato, editors. *Handbook of Memetic Algorithms*, volume 379 of *Studies in Computational Intelligence*. Springer Berlin, Heidelberg, Oct. 2011. doi: 10.1007/978-3-642-23247-3.

Y. Nesterov. *Introductory Lectures on Convex Optimization – A Basic Course*, volume 87 of *Applied Optimization*. Springer New York, NY, Dec. 2003. doi: 10.1007/978-1-4419-8853-9.

Q. U. Nguyen, E. Murphy, M. O'Neill, and X. H. Nguyen. Semantic-based subtree crossover applied to dynamic problems. In T. B. Ho, R. I. McKay, X. H. Nguyen, and T. D. Bui, editors, *The Third International Conference on Knowledge and Systems Engineering, KSE'2011*, pages 78–84, Hanoi University, Oct. 2011. IEEE. doi: 10.1109/KSE.2011.20.

Y. Niv. MUC1 and colorectal cancer pathophysiology considerations. *World Journal of Gastroenterology*, 14(14):2139–2141, 2008. doi: 10.3748/wjg.14.2139.

P. Nordin, W. Banzhaf, and F. D. Francone. Efficient evolution of machine code for CISC architectures using instruction blocks and homologous crossover. In L. Spector, W. B. Langdon, U.-M. O'Reilly, and P. J. Angeline, editors, *Advances in Genetic Programming 3*, chapter 12, pages 275–299. MIT Press, Cambridge, MA, USA, June 1999. ISBN 0-262-19423-6.

R. S. Olson, W. La Cava, P. Orzechowski, R. J. Urbanowicz, and J. H. Moore. PMLB: a large benchmark suite for machine learning evaluation and comparison. *BioData Mining*, 10(36):1–13, Dec. 2017. ISSN 1756-0381. doi: 10.1186/s13040-017-0154-4.

P. Orzechowski, W. La Cava, and J. H. Moore. Where are we now? A large benchmark study of recent symbolic regression methods. In *Proceedings of the Genetic and Evolutionary Computation Conference*, GECCO '18, pages 1183–1190. ACM, 2018. ISBN 9781450356183. doi: 10.1145/3205455.3205539.

N. Osman, N. O'Leary, E. Mulcahy, N. Barrett, F. Wallis, K. Hickey, and R. Gupta. Correlation of serum CA125 with stage, grade and survival of patients with epithelial ovarian cancer at a single centre. *Irish Medical Journal*, 101(8):245–247, Sep. 2008. ISSN 0332-3102.

R. K. Pace and O. W. Gilley. Using the spatial configuration of the data to improve estimation. *The Journal of Real Estate Finance and Economics*, 14 (3):333–340, May 1997. ISSN 1573-045X. doi: 10.1023/A:1007762613901.

L. Pagie and P. Hogeweg. Evolutionary Consequences of Coevolving Targets. *Evolutionary Computation*, 5(4):401–418, Dec. 1997. ISSN 1063-6560. doi: 10.1162/evco.1997.5.4.401.

Y. Pawitan. *In All Likelihood: Statistical Modelling and Inference using Likelihood*. Oxford University Press, 2001.

T. Pawlak, B. Wieloch, and K. Krawiec. Semantic backpropagation for designing search operators in genetic programming. *IEEE Transactions on Evolutionary Computation*, 19(3):326–340, June 2015. ISSN 1089-778X. doi: 10.1109/TEVC.2014.2321259.

T. P. Pawlak and K. Krawiec. Competent geometric semantic genetic programming for symbolic regression and Boolean function synthesis. *Evolutionary Computation*, 26(2):177–212, June 2018. doi: 10.1162/evco_a_00205.

F. Pedregosa, G. Varoquaux, A. Gramfort, V. Michel, B. Thirion, O. Grisel, M. Blondel, P. Prettenhofer, R. Weiss, V. Dubourg, et al. Scikit-learn: machine learning in python. *Journal of Machine Learning Research*, 12: 2825–2830, 2011.

B. K. Petersen, M. Landajuela, T. N. Mundhenk, C. P. Santiago, S. K. Kim, and J. T. Kim. Deep symbolic regression: recovering mathematical expressions from data via risk-seeking policy gradients. In *Proceedings of the International Conference on Learning Representations*, 2021.

D. N. Phong, N. Q. Uy, N. X. Hoai, and N. T. Thuy. Semantic based crossovers in tree-adjoining grammar guided genetic programming. In *IEEE International Conference on Computing and Communication Technologies, Research, Innovation, and Vision for the Future (RIVF 2013)*, pages 141–146, Nov. 2013. doi: 10.1109/RIVF.2013.6719883.

M. Pincus. Letter to the editor – a Monte Carlo method for the approximate solution of certain types of constrained optimization problems. *Operations Research*, 18(6):1225–1228, 1970. doi: 10.1287/opre.18.6.1225.

R. Poli. Discovery of symbolic, neuro-symbolic and neural networks with parallel distributed genetic programming. Technical Report CSRP-96-14, University of Birmingham, School of Computer Science, Aug. 1996. Presented at 3rd International Conference on Artificial Neural Networks and Genetic Algorithms, ICANNGA'97.

R. Poli, W. B. Langdon, and N. F. McPhee. *A Field Guide to Genetic Programming*. Lulu.com, 2008a. ISBN 978-1-4092-0073-4.

R. Poli, N. F. McPhee, and L. Vanneschi. The impact of population size on code growth in GP: analysis and empirical validation. In M. Keijzer, G. Antoniol, C. B. Congdon, K. Deb, B. Doerr, N. Hansen, J. H. Holmes, G. S. Hornby, D. Howard, J. Kennedy, S. Kumar, F. G. Lobo, J. F. Miller, J. Moore, F. Neumann, M. Pelikan, J. Pollack, K. Sastry, K. Stanley, A. Stoica, E.-G. Talbi, and I. Wegener, editors, *GECCO '08: Proceedings of the 10th Annual Conference on Genetic and Evolutionary Computation*, pages 1275–1282, Atlanta, GA, USA, July 2008b. ACM. doi: 10.1145/1389095.1389341.

W. H. Press, S. A. Teukolsky, W. T. Vetterling, and B. P. Flannery. *Numerical Recipes: The Art of Scientific Computing*. Cambridge University Press, 3rd edition, 2007. ISBN 0521880688.

M. Quade, M. Abel, K. Shafi, R. K. Niven, and B. R. Noack. Prediction of dynamical systems by symbolic regression. *Physical Review E*, 94(1), July 2016. doi: 10.1103/physreve.94.012214.

M. Quade, M. Abel, J. N. Kutz, and S. L. Brunton. Sparse identification of nonlinear dynamics for rapid model recovery. *Chaos: An Interdisciplinary Journal of Nonlinear Science*, 28(6), June 2018. doi: 10.1063/1.5027470.

M. Quade, J. Gout, and M. Abel. Glyph: symbolic regression tools. *Journal of Open Research Software*, 7(1):19, June 2019. doi: 10.5334/jors.192.

C. E. Rasmussen and C. K. I. Williams. *Gaussian Processes for Machine Learning*. The MIT Press, 2006. doi: 10.7551/mitpress/3206.001.0001.

I. Rechenberg. *Evolutionsstrategie: Optimierung technischer Systeme nach Prinzipien der biologischen Evolution*. PhD thesis, TU Berlin, 1971.

I. Rechenberg. *Evolutionsstrategie: Optimierung technischer Systeme nach Prinzipien der biologischen Evolution*. Number 15 in Problemata. Frommann-Holzboog, 1973.

J. Rissanen. Modeling by shortest data description. *Automatica*, 14(5):465–471, Sep. 1978. doi: 10.1016/0005-1098(78)90005-5.

P. Rockett. Pruning of genetic programming trees using permutation tests. *Evolutionary Intelligence*, 13(4):649–661, Dec. 2020. ISSN 1864-5917. doi: 10.1007/s12065-020-00379-8.

W. Roland, C. Marschik, M. Krieger, B. Löw-Baselli, and J. Miethlinger. Symbolic regression models for predicting viscous dissipation of three-dimensional non-newtonian flows in single-screw extruders. *Journal of Non-Newtonian Fluid Mechanics*, 268:12–29, June 2019. doi: 10.1016/j.jnnfm.2019.04.006.

D. G. Rosen, L. Wang, J. N. Atkinson, Y. Yu, K. H. Lu, E. P. Diamandis, I. Hellstrom, S. C. Mok, J. Liu, and R. C. Bast Jr. Potential markers that complement expression of CA125 in epithelial ovarian cancer. *Gynecologic Oncology*, 99(2):267–277, 2005.

F. Rothlauf. *Representations for Genetic and Evolutionary Algorithms*. Springer, 2nd edition, 2006. ISBN 3-540-25059-X.

F. Rothlauf. Representations for evolutionary algorithms. In *Proceedings of the 2016 on Genetic and Evolutionary Computation Conference Companion*, GECCO '16 Companion, pages 413–434. ACM, 2016. ISBN 9781450343237. doi: 10.1145/2908961.2926981.

C. Rudin. Stop explaining black box machine learning models for high stakes decisions and use interpretable models instead. *Nature Machine Intelligence*, 1(5):206–215, May 2019. ISSN 2522-5839. doi: 10.1038/s42256-019-0048-x.

C. Ryan, J. J. Collins, and M. O'Neill. Grammatical evolution: evolving programs for an arbitrary language. In W. Banzhaf, R. Poli, M. Schoenauer, and T. C. Fogarty, editors, *Proceedings of the First European Workshop on Genetic Programming*, volume 1391 of *Lecture Notes in Computer Sciences*, pages 83–96, Paris, Apr. 1998. Springer-Verlag. ISBN 3-540-64360-5. doi: 10.1007/BFb0055930.

B. Saha and K. Goebel. Battery data set. Technical report, NASA Prognostics Data Repository, NASA Ames Research Center, Moffett Field, CA, 2007. https://phm-datasets.s3.amazonaws.com/NASA/5.+Battery+Data+Set.zip, https://www.nasa.gov/content/prognostics-center-of-excellence-data-set-repository.

S. Sahoo, C. Lampert, and G. Martius. Learning equations for extrapolation and control. In J. Dy and A. Krause, editors, *Proceedings of the 35th International Conference on Machine Learning*, volume 80 of *Proceedings of Machine Learning Research*, pages 4442–4450. PMLR, July 2018.

A. K. Saini, L. Spector, and T. Helmuth. Environments with local scopes for modules in genetic programming. In *Proceedings of the Genetic and Evolutionary Computation Conference Companion*. ACM, July 2022. doi: 10.1145/3520304.3528958.

B. Schölkopf and A. J. Smola. *Learning with Kernels*. The MIT Press, 2018. doi: 10.7551/mitpress/4175.001.0001.

M. Schmidt and H. Lipson. Comparison of tree and graph encodings as function of problem complexity. In *Proceedings of the 9th Annual Conference on Genetic and Evolutionary Computation*, GECCO '07, pages 1674–1679, New York, NY, USA, 2007. ACM. ISBN 978-1-59593-697-4. doi: 10.1145/1276958. 1277288.

M. Schmidt and H. Lipson. Coevolution of fitness predictors. *IEEE Transactions on Evolutionary Computation*, 12(6):736–749, Dec. 2008a. ISSN 1941-0026, 1089-778X. doi: 10.1109/TEVC.2008.919006.

M. Schmidt and H. Lipson. Distilling free-form natural laws from experimental data. *Science*, 324(5923):81–85, Apr. 2009. doi: 10.1126/science.1165893.

M. Schmidt and H. Lipson. Age-fitness Pareto optimization. In R. Riolo, T. McConaghy, and E. Vladislavleva, editors, *Genetic Programming Theory and Practice VIII*, volume 8 of *Genetic and Evolutionary Computation*, chapter 8, pages 129–146. Springer, New York, NY, Ann Arbor, USA, May 2010. doi: 10.1007/978-1-4419-7747-2_8.

M. D. Schmidt and H. Lipson. Data-mining dynamical systems: automated symbolic system identification for exploratory analysis. In *Volume 2: Automotive Systems; Bioengineering and Biomedical Technology; Computational Mechanics; Controls; Dynamical Systems*. ASMEDC, Jan. 2008b. doi: 10.1115/esda2008-59309.

R. Schutt and C. O'Neil. *Doing Data Science: Straight Talk from the Frontline*. O'Reilly Media, Inc., 2013. ISBN 1449358659, 9781449358655.

H.-P. Schwefel. *Numerische Optimierung von Computer-Modellen mittels der Evolutionsstrategie*. Birkhäuser Basel, 1977. doi: 10.1007/978-3-0348-5927-1.

E. O. Scott and S. Luke. ECJ at 20: toward a general metaheuristics toolkit. In *Proceedings of the Genetic and Evolutionary Computation Conference Companion*, GECCO '19, pages 1391–1398. ACM, 2019. ISBN 9781450367486. doi: 10.1145/3319619.3326865.

D. P. Searson. Gptips 2: An open-source software platform for symbolic data mining. In A. H. Gandomi, A. H. Alvavi, and C. Ryan, editors, *Handbook of Genetic Programming Applications*, pages 551–573. Springer, Cham, 2015. ISBN 978-3-319-20883-1. doi: 10.1007/978-3-319-20883-1_22.

S. Silva, S. Dignum, and L. Vanneschi. Operator equalisation for bloat free genetic programming and a survey of bloat control methods. *Genetic Programming and Evolvable Machines*, 13(2):197–238, June 2012. ISSN 1389-2576. doi: 10.1007/s10710-011-9150-5.

G. Smits, A. Kordon, K. Vladislavleva, E. Jordaan, and M. Kotanchek. Variable selection in industrial datasets using Pareto genetic programming. In T. Yu, R. L. Riolo, and B. Worzel, editors, *Genetic Programming Theory and Practice III*, volume 9 of *Genetic Programming*, chapter 6, pages 79–92. Springer, Ann Arbor, May 2005. ISBN 0-387-28110-X. doi: 10.1007/0-387-28111-8_6.

T. Soule and J. A. Foster. Removal bias: a new cause of code growth in tree based evolutionary programming. In *1998 IEEE International Conference on Evolutionary Computation*, pages 781–786, Anchorage, Alaska, USA, May 1998. IEEE Press. ISBN 0-7803-4869-9. doi: 10.1109/ICEC.1998.700151.

L. Spector. Assessment of problem modality by differential performance of lexicase selection in genetic programming: a preliminary report. In K. McClymont and E. Keedwell, editors, *1st workshop on Understanding Problems (GECCO-UP)*, pages 401–408, Philadelphia, Pennsylvania, USA, July 2012. ACM. doi: 10.1145/2330784.2330846.

L. Spector and K. Stoffel. Ontogenetic programming. In J. R. Koza, D. E. Goldberg, D. B. Fogel, and R. L. Riolo, editors, *Genetic Programming 1996: Proceedings of the First Annual Conference*, pages 394–399, Stanford University, CA, USA, July 1996. MIT Press.

L. Spector, J. Klein, and M. Keijzer. The Push3 execution stack and the evolution of control. In *Proceedings of the 7th Annual Conference on Genetic and Evolutionary Computation*. ACM, June 2005. doi: 10.1145/1068009.1068292.

S. Stijven, W. Minnebo, and K. Vladislavleva. Separating the wheat from the chaff: on feature selection and feature importance in regression random forests and symbolic regression. In *Proceedings of the 13th Annual Conference Companion on Genetic and Evolutionary Computation*, GECCO '11, pages 623–630. ACM, 2011. ISBN 9781450306904. doi: 10.1145/2001858.2002059.

J. Stoer and R. Bulirsch. *Introduction to Numerical Analysis*. Springer New York, 2002. doi: 10.1007/978-0-387-21738-3.

M. Stone. Cross-validatory choice and assessment of statistical predictions. *Journal of the Royal Statistical Society: Series B (Methodological)*, 36(2): 111–133, 1974. doi: 10.1111/j.2517-6161.1974.tb00994.x.

M. Stone. Asymptotics for and against cross-validation. *Biometrika*, 64(1): 29–35, 1977. doi: 10.1093/biomet/64.1.29.

R. Storn and K. Price. Differential evolution – a simple and efficient heuristic for global optimization over continuous spaces. *Journal of Global Optimization*, 11(4):341–359, 1997. ISSN 0925-5001. doi: 10.1023/A:1008202821328.

D. R. Stoutemyer. Can the Eureqa symbolic regression program, computer algebra, and numerical analysis help each other? *Notices of the AMS*, 60(6): 713–724, June/July 2013.

W. A. Tackett and A. Carmi. The unique implications of brood selection for genetic programming. In *Proceedings of the 1994 IEEE World Congress on Computational Intelligence*, volume 1, pages 160–165, Orlando, Florida, USA, June 1994. IEEE Press. doi: 10.1109/ICEC.1994.350023.

I. M. Thompson, D. K. Pauler, P. J. Goodman, C. M. Tangen, M. S. Lucia, H. L. Parnes, L. M. Minasian, L. G. Ford, S. M. Lippman, E. D. Crawford, J. J. Crowley, and C. A. Coltman. Prevalence of prostate cancer among men with a prostate-specific antigen level ≤ 4.0 ng per milliliter. *New England Journal of Medicine*, 350(22):2239–2246, 2004. doi: 10.1056/NEJMoa031918. PMID: 15163773.

R. Tibshirani. Regression shrinkage and selection via the lasso. *Journal of the Royal Statistical Society. Series B (Methodological)*, pages 267–288, 1996.

A. Topchy and W. F. Punch. Faster genetic programming based on local gradient search of numeric leaf values. In *Proceedings of the 3rd Annual Conference on Genetic and Evolutionary Computation (GECCO'01)*, pages 155–162. Morgan Kaufmann San Francisco, CA, July 2001.

L. Trujillo, E. Z-Flores, P. S. Juárez-Smith, P. Legrand, S. Silva, M. Castelli, L. Vanneschi, O. Schütze, and L. Muñoz. Local search is underused in genetic programming. In R. Riolo, B. Worzel, B. Goldman, and B. Tozier, editors, *Genetic Programming Theory and Practice XIV*, Genetic and Evolutionary Computation, pages 119–137. Springer, Oct. 2018. doi: 10.1007/978-3-319-97088-2_8.

L. Trujillo, S. M. Winkler, S. Silva, and W. Banzhaf, editors. *Genetic Programming Theory and Practice XIX*. Springer Nature Singapore, 2023. doi: 10.1007/978-981-19-8460-0.

A. M. Turing. Computing machinery and intelligence. *Mind*, LIX(236):433–460, Oct. 1950. doi: 10.1093/mind/LIX.236.433.

S.-M. Udrescu, A. Tan, J. Feng, O. Neto, T. Wu, and M. Tegmark. AI Feynman 2.0: Pareto-optimal symbolic regression exploiting graph modularity. In H. Larochelle, M. Ranzato, R. Hadsell, M. Balcan, and H. Lin, editors, *Advances in Neural Information Processing Systems*, volume 33, pages 4860–4871. Curran Associates, Inc., 2020.

N. Q. Uy, N. X. Hoai, M. O'Neill, R. I. McKay, and E. Galvan-Lopez. Semantically-based crossover in genetic programming: application to real-valued symbolic regression. *Genetic Programming and Evolvable Machines*, 12(2):91–119, June 2011. ISSN 1389-2576. doi: 10.1007/s10710-010-9121-2.

N. Q. Uy, N. X. Hoai, M. O'Neill, R. I. McKay, and D. N. Phong. On the roles of semantic locality of crossover in genetic programming. *Information Sciences*, 235:195–213, June 2013. ISSN 0020-0255. doi: 10.1016/j.ins.2013.02.008.

L. Vanneschi, M. Castelli, and S. Silva. Measuring bloat, overfitting and functional complexity in genetic programming. In *Proceedings of the 12th Annual Conference on Genetic and Evolutionary Computation*. ACM, July 2010. doi: 10.1145/1830483.1830643.

L. Vanneschi, M. Castelli, L. Manzoni, and S. Silva. A new implementation of geometric semantic GP and its application to problems in pharmacokinetics. In *Proceedings of the 16th European Conference on Genetic Programming*, EuroGP'13, pages 205–216. Springer-Verlag, 2013. ISBN 9783642372063. doi: 10.1007/978-3-642-37207-0_18.

L. Vanneschi, M. Castelli, and S. Silva. A survey of semantic methods in genetic programming. *Genetic Programming and Evolvable Machines*, 15(2): 195–214, June 2014a. ISSN 1573-7632. doi: 10.1007/s10710-013-9210-0.

L. Vanneschi, S. Silva, M. Castelli, and L. Manzoni. Geometric semantic genetic programming for real life applications. In R. Riolo, J. H. Moore, and M. Kotanchek, editors, *Genetic Programming Theory and Practice XI*, pages 191–209. Springer New York, 2014b. ISBN 978-1-4939-0375-7. doi: 10.1007/978-1-4939-0375-7_11.

V. N. Vapnik. *Statistical Learning Theory*. Wiley, Sep. 1998. ISBN 978-0-471-03003-4.

J. H. Verner. Explicit Runge-Kutta methods with estimates of the local truncation error. *SIAM Journal on Numerical Analysis*, 15(4):772–790, 1978. ISSN 00361429.

D. Versino, A. Tonda, and C. A. Bronkhorst. Data driven modeling of plastic deformation. *Computer Methods in Applied Mechanics and Engineering*, 318: 981–1004, May 2017. doi: 10.1016/j.cma.2017.02.016.

M. Virgolin, T. Alderliesten, C. Witteveen, and P. A. N. Bosman. Scalable genetic programming by gene-pool optimal mixing and input-space entropy-based building-block learning. In *Proceedings of the Genetic and Evolutionary Computation Conference*, GECCO '17, pages 1041–1048. ACM, 2017. ISBN 9781450349208. doi: 10.1145/3071178.3071287.

M. Virgolin, T. Alderliesten, and P. A. N. Bosman. Linear scaling with and within semantic backpropagation-based genetic programming for symbolic regression. In *Proceedings of the Genetic and Evolutionary Computation Conference*, GECCO '19, pages 1084–1092. ACM, 2019. ISBN 9781450361118. doi: 10.1145/3321707.3321758.

E. J. Vladislavleva, G. F. Smits, and D. den Hertog. Order of nonlinearity as a complexity measure for models generated by symbolic regression via Pareto genetic programming. *IEEE Transactions on Evolutionary Computation*, 13 (2):333–349, Apr. 2009. ISSN 1089-778X. doi: 10.1109/TEVC.2008.926486.

J. von Neumann. *Theory of Self-Reproducing Automata*. University of Illinois Press, 1966.

H. Voss, M. Bünner, and M. Abel. Identification of continuous, spatiotemporal systems. *Physical Review E*, 57(3):2820–2823, Mar. 1998. doi: 10.1103/physreve.57.2820.

H. U. Voss, P. Kolodner, M. Abel, and J. Kurths. Amplitude equations from spatiotemporal binary-fluid convection data. *Physical Review Letters*, 83 (17):3422–3425, Oct. 1999. doi: 10.1103/physrevlett.83.3422.

G. P. Wagner and L. Altenberg. Perspective: complex adaptations and the evolution of evolvability. *Evolution*, 50(3):967–976, June 1996. doi: 10.2307/2410639.

S. Wagner, G. Kronberger, A. Beham, M. Kommenda, A. Scheibenpflug, E. Pitzer, S. Vonolfen, M. Kofler, S. Winkler, V. Dorfer, and M. Affenzeller. Architecture and design of the HeuristicLab optimization environment. In R. Klempous, J. Nikodem, W. Jacak, and Z. Chaczko, editors, *Advanced Methods and Applications in Computational Intelligence*, volume 6 of *Topics in Intelligent Engineering and Informatics*, pages 197–261. Springer, 2014. doi: 10.1007/978-3-319-01436-4_10.

C. S. Wallace and P. R. Freeman. Estimation and inference by compact coding. *Journal of the Royal Statistical Society Series B: Statistical Methodology*, 49 (3):240–252, 1987.

P. A. Whigham. Grammatically-based genetic programming. In J. P. Rosca, editor, *Proceedings of the Workshop on Genetic Programming: From Theory to Real-World Applications*, pages 33–41, Tahoe City, California, USA, July 1995.

B. Wieloch and K. Krawiec. Running programs backwards: instruction inversion for effective search in semantic spaces. In *Proceedings of the 15th Annual Conference on Genetic and Evolutionary Computation*, GECCO '13, pages 1013–1020. ACM, 2013. ISBN 9781450319638. doi: 10.1145/2463372.2463493.

S. Winkler, M. Affenzeller, W. Jacak, and H. Stekel. Classification of tumor marker values using heuristic data mining methods. In *Proceedings of the 12th Annual Genetic and Evolutionary Computation Conference, GECCO'10 – Companion Publication*, pages 1915–1922. ACM, July 2010. ISBN 9781450300735. doi: 10.1145/1830761.1830826.

S. M. Winkler. *Evolutionary System Identification: Modern Concepts and Practical Applications*. Number 59 in Johannes Kepler University, Linz, Reihe C. Trauner Verlag+Buchservice GmbH, 2009.

S. M. Winkler, M. Affenzeller, B. Burlacu, G. Kronberger, M. Kommenda, and P. Fleck. Similarity-based analysis of population dynamics in genetic programming performing symbolic regression. In R. Riolo, B. Worzel, B. Goldman, and B. Tozier, editors, *Genetic Programming Theory and Practice XIV*, pages 1–17, Ann Arbor, USA, May 2016. Springer. doi: 10.1007/978-3-319-97088-2_1.

I. H. Witten and E. Frank. *Data Mining: Practical Machine Learning Tools and Techniques*. Morgan Kaufmann, 2nd edition, 2005.

M. L. Wong and K. S. Leung. Evolutionary program induction directed by logic grammars. *Evolutionary Computation*, 5(2):143–180, June 1997. ISSN 1063-6560. doi: 10.1162/evco.1997.5.2.143.

P. Wong and M. Zhang. Algebraic simplification of genetic programs during evolution. Technical Report CS-TR-06-7, Computer Science, Victoria University of Wellington, New Zealand, Feb. 2006.

T. Worm and K. Chiu. Prioritized grammar enumeration: symbolic regression by dynamic programming. In *Proceedings of the 15th Annual Conference on Genetic and Evolutionary Computation*, pages 1021–1028. ACM, 2013.

I. W. Wright and E. J. Wegman. Isotonic, convex and related splines. *The Annals of Statistics*, 8(5), Sep. 1980. doi: 10.1214/aos/1176345140.

S. Wright. "Surfaces" of selective value. *Proceedings of the National Academy of Sciences*, 58(1):165–172, 1967. ISSN 0027-8424. doi: 10.1073/pnas.58.1.165.

H. Xie. *An Analysis of Selection in Genetic Programming*. PhD thesis, Victoria University of Wellington, 2009.

H. Xie and M. Zhang. Impacts of sampling strategies in tournament selection for genetic programming. *Soft Computing*, 16(4):615–633, Sep. 2011. doi: 10.1007/s00500-011-0760-x.

B. W. Yin, A. Dnistrian, and K. O. Lloyd. Ovarian cancer antigen CA125 is encoded by the MUC16 mucin gene. *International Journal of Cancer*, 98(5): 737–740, 2002. doi: 10.1002/ijc.10250.

E. Z-Flores, L. Trujillo, O. Schütze, and P. Legrand. Evaluating the effects of local search in genetic programming. In A.-A. Tantar, E. Tantar, J.-Q. Sun, W. Zhang, Q. Ding, O. Schütze, M. Emmerich, P. Legrand, P. Del Moral, and C. A. Coello Coello, editors, *EVOLVE - A Bridge between Probability, Set Oriented Numerics, and Evolutionary Computation V*, pages 213–228. Springer International Publishing, 2014. ISBN 978-3-319-07494-8.

B.-T. Zhang and H. Mühlenbein. Balancing accuracy and parsimony in genetic programming. *Evolutionary Computation*, 3(1):17–38, 1995. doi: 10.1162/evco.1995.3.1.17.

M. Zhang, P. Wong, and D. Qian. Online program simplification in genetic programming. In T.-D. Wang, X. Li, S.-H. Chen, X. Wang, H. A. Abbass, H. Iba, G. Chen, and X. Yao, editors, *Simulated Evolution and Learning, Proceedings 6th International Conference, SEAL 2006*, volume 4247 of *Lecture Notes in Computer Science*, pages 592–600, Hefei, China, Oct. 2006. Springer. ISBN 3-540-47331-9. doi: 10.1007/11903697_75.

D. G. Zill and M. R. Cullen. *Advanced Engineering Mathematics*. Jones and Bartlett Publishers, 3rd edition, 2005. ISBN 978-0-7637-4591-2.

E. Zitzler and L. Thiele. Multiobjective evolutionary algorithms: a comparative case study and the strength Pareto approach. *IEEE Transactions on Evolutionary Computation*, 3(4):257–271, 1999. doi: 10.1109/4235.797969.

E. Zitzler, M. Laumanns, and L. Thiele. SPEA2: improving the strength Pareto evolutionary algorithm. Technical Report TIK Report 103, Computer Engineering and Networks Laboratory (TIK), Swiss Federal Institute of Technology (ETH) Zurich, May 2001.

H. Zou and T. Hastie. Regularization and variable selection via the elastic net. *Journal of the Royal Statistical Society: Series B (Statistical Methodology)*, 67(2):301–320, 2005.

Index